HANDBOOK OF ALGAE

Handbook
of
ALGAE

JAGDISH ACHARYA

RAJAT PUBLICATIONS
NEW DELHI - 110 002 (INDIA)

RAJAT PUBLICATIONS
4675/21, Ansari Road, Daryaganj
New Delhi- 110 002 (India)
Phones : 23267924, 22507277
E-mail: rajat_publication@yahoo.com

Handbook of Algae

Edition 2026

ISBN 978-81-7880-382-1

PRINTED IN INDIA

Published by Mrs. Seema Wasan for Rajat Publications, New Delhi and Printed atTarun Offset Press, Delhi.

Preface

Today, the algae have taken on added economic importance (as food, for commercial products, water contamination, toxin-production, sewage oxidation, etc.). Further, with the development of refined Culture Techniques and the Electron Microscope, Physiologists and cytologists are employing algae more than ever as subjects of study for fundamental scientific information. The Geneticist, Limnologist, Palaeontologist, and plant evolutionist find investigations of algae to be both necessary and highly rewarding. One needs but to consider the current flood of literature dealing with the many aspects of algae (Taxonomic, Physiological, Morphological, Ontological) to gain an appreciation of the place that algae hold in human and scientific affairs.

"This book has been prepared, for both the specialist and especially for the beginning student, because of this burgeoning interest in algae. Hopefully, those who are learning to identify algae for one reason or another will find pertinent information in the pages that follow. Emphasis has also been given to the relationship between algae and water quality and conservation. Finally, the information on life histories and ecology, and their relationship to algal form and structure, should be of interest to the student of algae morphology.

JAGDISH ACHARYA

Contents

Preface ..... *vii*

1 Introduction 1

2 Cyanophyta 13

3 Chlorophyta 26

4 Euglenophyta 62

5 Phaeophyta 76

6 Rhodophyta 105

7 Cryptophyta Pyrrhophyta 121

8 Chrysophyta Bacillariophyta 131

9 Fossil Algae 174

10 Soil Algae and Symbiosis 192

11 Algal Physiology 200

12 Ecology of Algae 217

13 Reproduction in Algae 253

14 Geographical Distribution of Algae 266

1
Introduction

Algae, diverse group of simple, plantlike organisms. Like plants, most algae use the energy of sunlight to make their own food, a process called photosynthesis. However, algae lack the roots, leaves, and other structures typical of true plants. Algae are the most important photosynthesizing organisms on Earth. They capture more of the sun's energy and produce more oxygen (a byproduct of photosynthesis) than all plants combined. Algae form the foundation of most aquatic food webs, which support an abundance of animals.

Algae vary greatly in size and grow in many diverse habitats. Microscopic algae, called phytoplankton, float or swim in lakes and oceans. Phytoplankton are so small that 1000 individuals could fit on the head of a pin. The largest forms of algae are seaweeds that stretch 100 m from the ocean bottom to the water's surface. Although most algae grow in fresh water or seawater, they also grow on soil, trees, and animals, and even under or inside porous rocks, such as sandstone and limestone. Algae tolerate a wide range of temperatures and can be found growing in hot springs, on snow banks, or deep within polar ice.

Algae also form mutually beneficial partnerships with other organisms. Algae live with fungi to form lichens—plantlike or branching growths that form on boulders, cliffs, and tree trunks. Algae called *zooxanthellae* live inside the cells of reef-building coral. In both cases, the algae provide oxygen and complex nutrients to their partner, and in return they receive protection and simple nutrients. This arrangement enables both partners to survive in conditions that they could not endure alone.

The earliest life-forms on this planet are thought to be early ancestors of cyanobacteria, a type of algae formerly called blue-green algae. Fossilized cyanobacteria have been found in rocks more than 3 billion years old. These early algae formed when there was no oxygen in the atmosphere, and scientists theorize that as the algae photosynthesized, they released oxygen as a byproduct, which eventually accumulated in the atmosphere. Algae were probably the first organisms capable of photosynthesis and, until the appearance of plants on earth, the only photosynthesizers for billions of years.

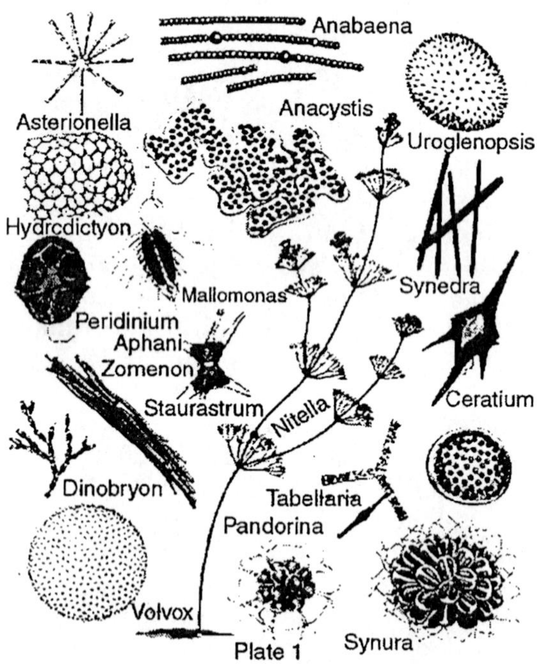

Fig. Various Types of Algae.

PHYSICAL CHARACTERISTICS

With the exception of the cyanobacteria, algae are eukaryotes—that is, the insides of their cells are organized into separate membrane-wrapped organelles, including a nucleus and mitochondria. An important organelle found in eukaryotic algae is the chloroplast, which contains the light-absorbing pigments responsible for capturing the energy in sunlight during photosynthesis. In most algae the primary pigment is chlorophyll, the same green pigment used in plants.

Many algae also contain secondary pigments, including the carotenoids, which are brown or yellow, and the phycobilins, which are red or blue. Secondary pigments give algae their colourful hues.

The cyanobacteria are prokaryotes—that is, relatively simple unicellular organisms lacking a nucleus and other membrane-bound organelles. As their modern name implies, the cyanobacteria have many characteristics that resemble bacteria.

Like plants, most algae have rigid cell walls composed largely of cellulose. An exception is the diatom, whose cell wall is composed primarily of silica, which provides rigidity and also produces elaborately sculpted patterns of grooves that serve as identifying features. Many eukaryotic algae have whiplike appendages called flagella attached to their cell walls. By beating flagella in a rotary movement, these algae are able to move through water with considerable speed. A few algae that are devoid of rigid cell walls are able to protrude one part of the body ahead of the other to crawl on solid surfaces in an amoeba-like fashion.

Algae come in a variety of shapes and forms. The simplest form is the single, self-sufficient cell, such as *Euglena*, dependent only on sunlight and carbon dioxide and minerals from the water. Numerous one-celled algae may clump together to form a colony. Although these cells are attached to one another, each cell within a colony continues to function independently. Still other algae are multicellular organisms. In the simplest multicellular algae, the cells are joined end to end, forming filaments, both branched and unbranched. More complex structures may be shaped like a small disc, tube, club, or even a tree. The most complex algae have highly specialized cells. Some seaweeds, for instance, have a variety of specialized tissues, including a rootlike *holdfast*, a *stipe*, which resembles a plant stalk, and a leaflike *blade*.

While most algae create their own food through photosynthesis, some are unable to photosynthesize. These algae ingest food from external sources by absorbing simple nutrients through the cell membrane. To absorb more complex nutrients, algae that lack rigid walls are able to engulf food particles and digest them.

Some of the algae known as dinoflagellates extend a feeding tube, called a peduncle, to suck in food. Other dinoflagellates use special harpoonlike structures to snare their food. Some algae are parasites, living in or on another organism from which they get their food. Some parasitic red algae live off other red algae, and parasitic dinoflagellates live in the intestines of some marine animals, such as copepods and annelids.

REPRODUCTION

Algae reproduce in astoundingly diverse ways. Some reproduce asexually, others use sexual reproduction, and many use both. In asexual reproduction an individual reproduces without combining its genetic material with that from another individual. The simplest form of asexual reproduction is binary fission, in which a unicellular organism simply divides into two new individuals.

Some multicellular algae, including *Sargassum*, reproduce asexually through fragmentation, in which fragments of the parent develop into new individuals. In a similar process called budding, special buds detach from multicellular algae and develop into new individuals, commonly found in *Sphacelaria*. Many algae produce special cells called spores that are capable of growing into new individuals. If these spores move about using flagella, they are known as zoospores.

In sexual reproduction, genetic material from two individuals is combined. The simplest form of sexual reproduction in algae is conjugation, in which two similar organisms fuse, exchange genetic material, and then break apart. In *Spirogyra*, which produces both asexually and sexually, two long, unbranched filaments join via conjugation tubes, through which genetic material is exchanged between cells. Most multicellular algae undergo a more complex form of sexual reproduction involving the union of special reproductive cells, called gametes, to form a single cell, known as a zygote.

Many algae incorporate both sexual and asexual modes of reproduction. This is well demonstrated in the life cycle of the alga *Chlamydomonas*. The mature alga is a single haploid cell—that is, it contains only one set of chromosomes. During asexual reproduction the cell undergoes mitosis, a type of cell division that produces genetically identical offspring. Four daughter cells are created that emerge from the enclosing parent cell as spores. The spores develop into mature haploid cells that are genetically identical to the parent cell.

Certain environmental conditions, such as lack of nutrients or moisture, may trigger the haploid daughter cells to undergo sexual reproduction. Instead of forming into spores, the haploid daughter cells form gametes that have two different mating strains. These two strains are structurally similar and are called plus and minus strains.

Opposite mating strains fuse in a process known as isogamy to form a diploid zygote, which contains two sets of chromosomes. After a period of dormancy, the zygote undergoes meiosis, a type of cell division that reduces the genetic content of a cell by half. This cell division produces four genetically unique haploid cells that eventually grow into mature cells.

Some multicellular green algae, such as *Ulva*, follow a distinct pattern of reproduction called alternation of generations, in which it takes two generations—one that reproduces sexually and one that reproduces asexually—to complete the life cycle. The two mature forms of the algae, alternating between diploid and haploid individuals, are identical in appearance, or isomorphic. The haploid form, called a gametophyte, undergoes mitosis to produce haploid gametes. These gametes unite to form a diploid zygote, which develops into the diploid form called a sporophyte. The sporophyte undergoes meiosis to form haploid spores that, in turn, form gametophytes.

Not all algae that undergo alternation of generations have haploid and diploid forms that look alike. In the life cycle of the seaweed *Laminaria*, the gametophyte and the sporophyte are distinct in appearance, or heteromorphic. The *Laminaria* sporophyte appears as long, bladelike structures that grow on rocks just below the water in intertidal zones. The gametophyte is short, with branched filaments.

TYPES OF ALGAE

The most common classification system distributes algae

in more than one kingdom. Most algae are classified in the Kingdom Protista, along with other eukaryotic organisms that lack true specialized tissues. The cyanobacteria, however, are classified with the bacteria in the Kingdom Prokaryotae, which consists of prokaryotic organisms.

This classification system continues to be intensely debated as new research increases our understanding of the way that these organisms are related. The classification scheme proposed in 1997 that divides algae, including the cyanobacteria, into 11 different phyla.

GREEN ALGAE

Green algae form the phylum Chlorophyta and are named for their green chloroplasts, which are similar in composition to the chloroplasts found in land plants. Green algae range in shape from unicellular plankton that grow in lakes and oceans to colonial filaments of pond scum to leaflike seaweeds that grow along rocky and sandy intertidal areas.

Some green algae also live on tree trunks and soil. Several green algal species are symbiotic, forming lichens with fungi or living with corals. Green algae may also be found inside freshwater sponges, giving the sponges a bright green colour, and in permanent snow banks, where a secondary pigment masks the chlorophyll and turns the snow a reddish colour.

More than 500 genera and 8000 species of green algae have been identified. Some familiar green algae include the genus *Spirogyra*, known for its spiral-shaped chloroplasts, and the desmids, recognized by their characteristic shape—two symmetrical halves, joined by a small bridge. The green algae known as Stoneworts often grow several feet in length. Their name comes from calcium crusts that make them feel like stone.

Most green algae reproduce both sexually and asexually. Alternation of generations, where the algae alternates between gametophyte and sporophyte generations, is common among the multicellular green algae.

RED ALGAE

Red algae form the phylum Rhodophyta with approximately 500 genera and 6000 species. Found in warm coastal waters and in water as deep as 260 m (850 ft), red algae species adapt to varied water depths by having different proportions of pigments. Their red colour is due to a red pigment, phycoerythrin, which is well suited to absorb the blue light that penetrates deeper into water than the other colours of light. Red algae found in deep water may be almost black due to a high concentration of phycoerythrin.

At moderate depths red algae appear red, while in shallow water they may appear green because a smaller proportion of phycoerythrin is unable to mask the green of chlorophyll. Most red algae are multicellular and come in a variety of shapes, including filaments, which are shaped like a blade of grass, and seaweed shapes. Unlike most other eukaryotic algae, red algae have no flagella.

Red algae use diverse strategies to reproduce, including fragmentation and spore production. One unusual strategy, found in many species including those in the genus *Polysiphonia,* involves the alternation among three generations. A diploid sporophyte produces diploid spores that germinate into another diploid sporophyte that looks completely different from the first one. Meiosis occurs in the second sporophyte, producing haploid spores that germinate into gametophytes. Surprisingly, in some species, the gametophytes look nearly identical to the second sporophyte.

Almost all red algae live in marine habitats, although some species are found in fresh water or damp soil. Many types of seaweed are red algae, typically found growing along the coast and attaching firmly to the seafloor using a rootlike holdfast. In some species, called coralline algae, the cell walls become hardened with calcium carbonate. Coralline algae are important members of coral reefs, producing new material and cementing together other organisms.

GOLDEN-BROWN ALGAE, BROWN ALGAE AND DIATOMS

Golden-brown algae, brown algae, and diatoms form the large and complex phylum Heterokontophyta, with organisms ranging in size from a fraction of a millimeter to more than 100 m (300 ft) long. Heterokontophyta have carotenoid secondary pigments that tend to mask the green of the primary chlorophyll pigment, giving them a golden or golden-brown appearance. Flagellated cells in this phylum have two types of flagella: One is smooth, while the other has two rows of stiff hairs running down opposite sides of the flagellum. Algae in this phylum typically have an eyespot that can detect light.

The golden-brown algae, also known as the yellow-brown algae, include about 200 genera and 1000 species that receive their characteristic colouring from the carotenoid pigment fucoxanthin. These algae are mostly unicellular or colonial, swimming or floating in lakes and oceans as phytoplankton. In shallow ponds that dry up in summer or freeze completely in winter, golden-brown algae survive by forming protective cysts that can withstand the harsh conditions. When favorable conditions return, the algae emerge from the cysts. Like so many other algae, the unicellular algae tend to reproduce through fission, while the multicellular and colonial forms reproduce either through fragmentation or through spore production.

Diatoms are best known for their glasslike cell wall made of silica. The cell wall has ornate ridge patterns. A diatom consists of two overlapping halves that fit together like a shoebox or a petri dish, with the lid slightly larger and fitting over the base. During asexual cell division, the two glass walls separate and serve as the lids for two new glass bases.

The new diatom that grew from the lid is the same size as the parent diatom, while the diatom that grew from the smaller base is slightly smaller than its parent. Sexual reproduction occurs when the succeeding generations shrink to a critical size. These smallest diatoms form gametes that shed their glass walls. Upon fertilization, the zygotes absorb water to swell and then secrete new, full-sized silica coverings.

A very large class with more than 250 genera and 8000 species, diatoms are found floating in freshwater and seawater, growing attached to the seafloor, or growing on soil. The cells are either unicellular or form colonial chains of round cells. When an organism dies, its silica cell wall remains intact.

Over time these shells have accumulated to form layers of soft rock in some geologic formations.

This diatomaceous earth is mined and quarried for use in filters and bleaching agents, as an abrasive powder for cleaning and polishing metals, and for insect pest control (the broken cell walls of silica tear the insect gut).

Brown algae include over 260 genera and 1500 species. Multicellular algae, they may range from tiny filaments to the largest and most complex algae, such as the kelps, with leaflike blades and stems that can be up to 100 m (300 ft) long. Most brown algae grow in marine waters near the coast, attached to rocks either along the shoreline or underneath the ocean surface. One type, *Sargassum,* forms huge floating masses in the middle of the Sargasso Sea. The brown or olive colour is due to the pigment fucoxanthin. The life cycles of brown algae vary considerably, but most demonstrate alternation of generations.

DINOFLAGELLATES

Dinoflagellates of the phylum Dinoflagellata are mostly unicellular organisms that may be covered with stiff cellulose plates that resemble armored helmets. Many species have unusual ornamentation, such as horns, spines, or wings. A narrow groove encircles the armor, and a second groove runs perpendicular to the first groove. Flagella beat within these grooves, causing the dinoflagellates to spin like tops as they move through the water. Most of the 130 genera and 2000 species in this phylum are planktonic and live in saltwater, although there are many freshwater planktonic representatives as well.

Many dinoflagellate species lack chloroplasts and are dependent on other species for their food. Some are parasites, but most are carnivores, using special harpoonlike structures called trichocysts to capture other organisms to eat. In contrast, several of the photosynthetic species live inside the tissues of invertebrate animals, such as corals and giant clams. These dinoflagellates share the food they photosynthesize with their host, and in return, receive protection and some nutrients.

Under favorable environmental conditions, some

dinoflagellate species experience population explosions, known as blooms. If the species involved in the bloom have red pigments, their concentration can be high enough to turn the seawater red, forming red tides. Dinoflagellate blooms can be quite destructive. During the night when photosynthesis halts, such a high concentration of individuals can deplete the oxygen in the water, suffocating fish.

Some dinoflagellates release toxins, some of which kill fish, while other toxins are passed up the food chain until they reach humans, where they can cause paralytic shellfish poisoning and ciguatera fish poisoning. Recently, the dinoflagellate *Pfiesteria piscicida* has caused fish, shellfish, and human disease in estuaries of the southeastern United States.

CYANOBACTERIA

Unlike other algae, the cyanobacteria are prokaryotes—single-celled organisms with characteristics that cause biologists to debate whether they are really algae or bacteria. Cyanobacteria are found nearly everywhere, occurring in typical aquatic and terrestrial habitats as well as in such extreme sites as hot springs with temperatures as high as 71° C (160° F) and crevices of desert rocks. Cyanobacteria make up the phylum Cyanophyta, and this phylum contains about 150 genera and 2000 species worldwide.

Like other bacteria, cyanobacteria do not have organelles such as nuclei, mitochondria, or chloroplasts. Cyanobacteria are distinguished from bacteria by the presence of internal membranes, called thylakoids, that contain chlorophyll and other structures involved in photosynthesis. While higher plants have two kinds of chlorophyll, called a and b, cyanobacteria contain only chlorophyll a. Cyanobacteria colour varies from blue-green to red or purple and is determined by the proportions of two secondary pigments, c-phycocyanin (blue) and c-phycoerythrin (red), which tend to mask the green chlorophyll present in the thylakoids.

Cyanobacteria reproduce asexually by binary fission, spore production, or fragmentation, forming singular cells, colonies, filaments, or gelatinous masses. Although most lack flagella and are nonmotile, filamentous forms such as

Oscillatoria rotate in a screwlike manner, and the gelatinous forms glide along their slimy mucus.

Cyanobacteria may be both beneficial and harmful to humans: Some act as natural fertilizers in some habitats, especially rice paddies, whereas others produce toxins. Mild cyanobacteria toxins in lakes and oceans cause a rash known as swimmer's itch, while powerful neuromuscular toxins released by other cyanobacteria can kill fish living in the water or the animals that drink the water. In certain conditions, cyanobacteria may form dense blooms, which may produce toxins that make seafood poisonous to humans. Even if the cyanobacteria do not produce toxins, blooms can cause water to have an unpleasant taste and odor.

ALGAE USES

Human ingenuity has found many uses for algae. Algae provide food for people and livestock, serve as thickening agents in ice cream and shampoo, and are used as drugs to ward off diseases. More than 150 species of algae are commercially important food sources, and over $2 billion of seaweed is consumed each year by humans, mostly in Japan, China, and Korea.

The red alga *Porphyra*, called nori, is the most popular food product. After harvesting, nori is dried, pressed into sheets, and used in soups, sauces, sushi, and condiments. Algae are considered nutritious because of their high protein content and high concentrations of minerals, trace elements, and vitamins. The high iodine content of many edible algae may contribute to the low rates of goiter observed in countries where people frequently eat algae.

In coastal areas of North America and Europe, seaweeds are fed to farm animals as a food supplement. Cyanobacteria species that are high in protein, such as *Spirulina*, are grown commercially in ponds and used mostly as a health food and cattle dietary supplement. Seaweeds also are applied to soils as a fertilizer and soil conditioner, as their high concentrations of potassium and trace elements improve crop production.

Some species of cyanobacteria can turn atmospheric nitrogen into ammonia, a form that can then be used by plants as a nutrient. Farmers in tropical countries grow cyanobacteria in their flooded rice paddies to provide more nitrogen to the rice, increasing productivity as much as tenfold.

Seaweeds are a critical source of three chemical extracts used extensively in the food, pharmaceutical, textile, and cosmetic industries. Brown algae yield alginic acid, which is used to stabilize emulsions and suspensions; it is found in products such as syrup, ice cream, and paint.

Different species of red algae provide agar and carrageenan, which are used for the preparation of various gels used in scientific research. Bacteria, fungi, and cell cultures are commonly grown on agar gels. Agar is also used in the food industry to stabilize pie fillings and preserve canned meat and fish. Carrageenan is also used as a thickening and stabilizing agent in products such as puddings, syrups, and shampoos.

Algae have been used for centuries, especially in Asian countries, for their purported powers to cure or prevent illnesses as varied as cough, gout, gallstones, goiter, hypertension, and diarrhea. Recently, algae have been surveyed for anticancer compounds, with several cyanobacteria appearing to contain promising candidates. Diatoms also have been used in forensic medicine, as their presence in the lungs can indicate a person died due to drowning.

Algae can also serve as indicators of environmental problems in aquatic ecosystems. Because algae grow quickly and are sensitive to changing environmental conditions, they are often among the first organisms to respond to changes. Depletion of the diatom community in the Florida Everglades provided strong evidence of phosphorus-related changes in this unique ecosystem. Algal blooms may deplete oxygen concentrations in water and smother fish and plant life, as well as prevent light penetration for algae at lower depths, preventing photosynthesis.

2
Cyanophyta

CYANOPHYCEAE

The name Cyanophyceae has recently been generally adopted for this class previously known as the Myxophyceae. Apart from the main feature that the cells lack true nuclei, they show other features which, with the bacteria, distinguish themfrom the Eucaryota. The cell envelope in the Cyanophyceae consists of an outer sheath (capsule in the bacteria), an outer membrane and an inner one and a plasma membrane. The cell envelope is composed largely of mucopeptides built up from amino sugars (e.g. glucosamine) and amino acids (e.g. muramic, diaminopimelic), whereas the cell wall of other algae lack such amino constituents.

The bacteria and Cyanophyceae lack mitochondria, true vacuoles and endoplasmic reticula. Sterols have not been detected in bacteria, and the absence of sterolsin the Cyanophyceae was also an accepted fact. Recent work has shown the presence of cholesterol and 8-sitosterol in *Anacystis* and *Fremyella*. There is no membrane bounded chloroplast and in the blue-green algae the photosynthetic lamellae are usually distributed in the peripheral cytoplasm.

The separate microscopic plants are characterized by a bluish-green colour, occasionally red, which varies greatly in shade, depending upon the relative proportions of chlorophyll #, carotenoids, phycocyanin and phycoerythrin. As well as the photosynthetic lamellae the Cyanophyceae possess small granules resembling ribosomes. The photosynthetic lamellae can be arranged irregularly or in parallel stacks, but they are always separated by a layer of cytoplasm. Fine fibrils of DNA (deoxyribose nucleic acid) are a feature of the nucleoplasm and these are either distributed throughout the cell or concentrated in the central portion.

The use of the electron microscope has shown that the former idea of an outer chromatic zone and an inner colourless zone is not valid and that cells of different species have the structural units just described, which can be arranged in a number of different ways.

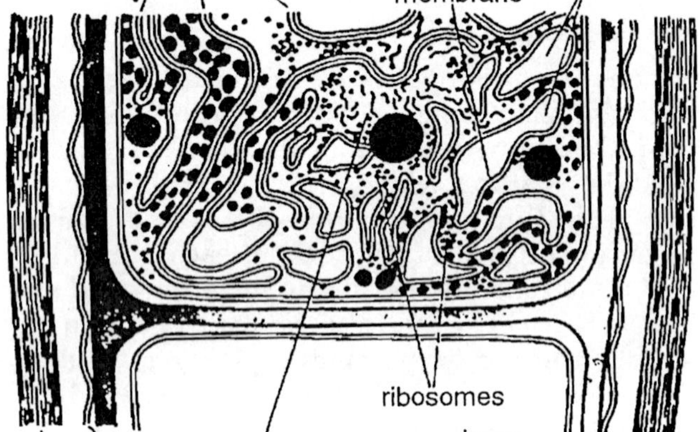

Fig. Diagram of Cell of Calothrix

The protoplast is normally devoid of vacuoles, except in old cells, and this is a major factor contributing to the great resistance of the plants to desiccation and of the cells to plasmolysis. In some forms (e.g. *Anabaena, Oscillatorid),* and principally in species which are planktonic, pseudovacuoles (gas vacuoles) may be found and it is supposed that these contribute towards their buoyancy by virtue of the gas that they contain. In *Microcystis* these consist of a single layer of particles, a few nanometers in dimension, linearly arranged around the axis of the vacuole which has approximate dimensions of 70 x 400 nm. The mucilaginous cell sheath, which may surround the whole cell, e.g. *Anacystis,* or form a cylindrical sheath, e.g. *Oscillatoria,* or an interrupted sheath, e.g. *Anabaena,* appears to be composed of cellulose fibrils.

There is considerable variation in the composition of the

different cell sheaths, and the amount of material laid down frequently depends upon the external environment. Colour of the sheaths is modified by acidity of the medium and also by the action of fungal parasites. Protoplasmic connections between mature cells are known in one order, the Stigonematales.

The class is characterized by a general absence of well-marked reproductive organs, sexual reproduction or motile reproductive bodies. Fusion of filaments has been reported for a strain of *Nostoc muscorum*.

The coccoid forms (spherical cells) multiply by cell division which takes place by means of a progressive constriction, whilst in some other types the cell contents give rise to a number of non-motile bodies that are termed endospores (formerly called gonidia).

All stages from simple binary fission to endospores can be found:

1. Binary fission, e.g. *Chroococcus turgidus*
2. Quadrants and octants formed, e.g. *Anacystis montana*
3. Numerous small daughter cells are produced in which there is retention of individual sheaths
4. The same without individual sheaths
5. Abstricted endospores, e.g. *Entophysalis*.

During cell division and at binary fission the photosynthetic lamellae become invaginated at the point of division and a new membrane is formed here by invagination of the inner membrane of the long wall.

Many of the filamentous forms produce specialized cells known as *hetero-cysts*. These are enlarged cells which possess thickened walls, particularly at their poles, and they usually occur singly, though occasionally they may be formed in rows. They develop from an ordinary vegetative cell, but during development they remain in protoplasmic communication with neighbouring cells by a small cytoplasmic connection.

During development the hjeterocyst increases in size, and there is an apparent increase in the cytoplasmic matrix, disorganization of the photosynthetic lamellae system and disappearance of some of the cytoplasmic inclusions. As the heterocyst enlarges a thick outer envelope is laid down outside the cell. There is usually a pore at one end of the terminal heterocyst (at both ends if intercalary) at which the disorganized chloroplast lamellae often reaggregate.

There are connections to plasma membranes of adjacent cells, but there does not appear to be any direct protoplasmic connection. It has been shown that heterocyst formation occurs only when the nitrogen level falls below a certain level and that heterocysts are incapable of photosynthesis but may fix nitrogen.

The widespread occurrence of heterocysts, especially terminal ones, and their specialized structure indicates that they must fulfil a significant role. Various suggestions have been made as to their function, and in some cases they do seem to determine the breaking up of the trichomes or threads into hormogones. These hormogones are short lengths of thread which are cut off. The trichome plus the sheath is termed a filament.

In the past it has been suggested that the heterocysts represent archaic reproductive organs that are now functionless. Fritsch suggested that the heterocysts, during the vegetative period, secrete substances that stimulate growth and cell division. He also suggested that at the time of reproduction the nature of the secretion changes and as a result akinetes are formed.

Recently Wolk has demonstrated again that under certain conditions (e.g. transfer to ammonium chloride and glucose) heterocysts will 'germinate' to give rise to multicellul argermlings. He also found that sporulation was stimulated by phosphate starvation and that spores could germinate on phosphate agar. Wolk then suggested that these two adaptations are survival mechanisms for deficiencies of nitrogen and phosphate. In other experiments he has shown that sporulation of akinetes adjacent to heterocysts is dependent upon the heterocysts.

Hormogones, besides being cut off by heterocysts, may also be produced by the development of biconcave 'separation discs' at intervals along the filament, or they may break off from the ends of the trichomes. The so-called separation discs are formed by the solidifying of a cell and its contents in the filament. The hormogones, together with certain of the filamentous types, e.g. *Oscillatoria*, and a few Chroococcales, exhibit a slow forward motion and associated with this motion is an active and continual secretion of mucilage along the sides of the filaments. The actual mechanism said to be responsible for locomotion is rhythmic longitudinal waves passing from end to end of the trichomes and caused by changes in the volumes of the protoplasts.

This is probably an over simplification and until further experimental work has been carried out interpretation should be reserved. In four species of *Phormidium* that have been fairly thoroughly studied, it has been found that movement is a response to light, and that it occurs between light intensities of 0-02 lx and 30,000 lx with optimum response at 2,000 lx. It is suggested that there are three different light reaction systems that are independent of each other, so that there is some degree of complexity. The principal system is related to chlorophyll alone but there is another based upon both chlorophyll and the biliproteins. A third system is related to a photo-receptor of unknown character.

Yet another type of structure, the thick-walled resting spore or akinete, occurs in many of the filamentous forms belonging to the Nostocaceae and Rivulariaceae, and these akinetes normally develop next to a heterocyst, either singly or in a series. On germination they give rise to a new filament.

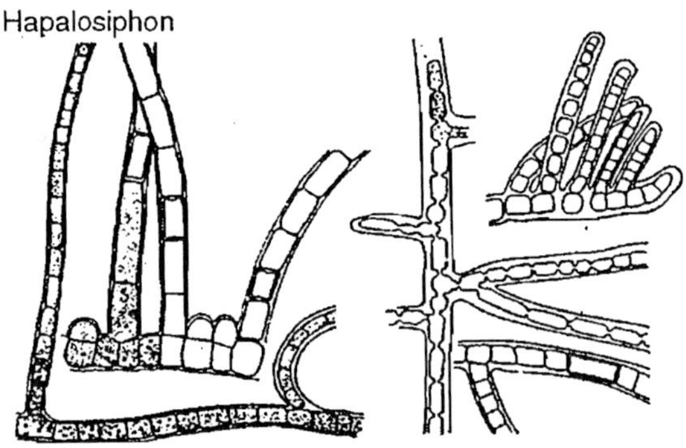

Fig. Hapalosiphon types found in all Mastigocladus Plants.

Many of the forms aggregate into colonies, but in some of the Chroococca-ceae the macroscopic plant mass represents an association of such colonies rather than a single colony. The form which any colony may take depends on first, the planes of cell division, and second, the effect of environment, which may determine the consistency of the mucilage. Uneven temperatures, sometimes produce irregular growth.

It has been shown experimentally that the environment may affect the shape of colonies of *Microcystis* and *Chroococcus turgidus* and determines the size of *Rivularia haematites.*

The single floating unicell may not necessarily represent the most primitive condition. The class is exceptionally ancient and must have evolved at a very early stage. It is known that the early waters of the world were not saline as at present and they were probably very much warmer and the fact that many Cyanophyceae are found associated with hot springs may be of significance. Schwabe has suggested that it is in such places the most primitive forms may be found, and he proposed a radiating system based upon the behaviour of *Mastigocladus laminosus,* a wide-spread thermal alga.

In this alga, under different conditions, a wide range of morphological form can develop. This plasticity is regarded as characteristic of a primitive thermal alga and various lines of morphological specialization are proposed as follows:

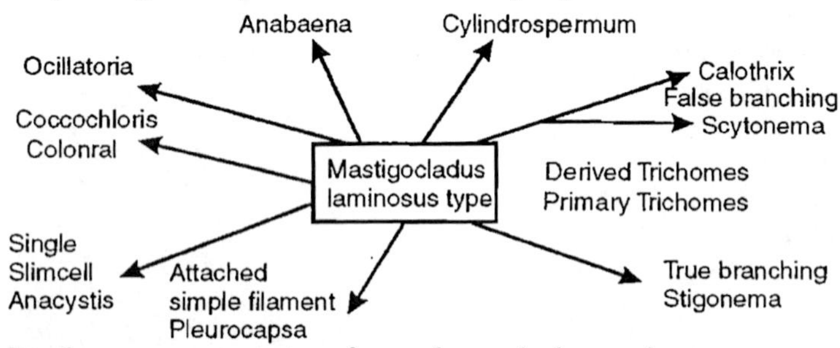

In the present state of our knowledge other courses of evolutionary development could be envisaged. Further intensive studies with the electron microscope may provide some additional data, though in view of the comparative simplicity of cellular structure, this is not very likely.

As may be expected from a primitive and ancient group there is evidence of parallel development when compared with plants from other groups, and this follows what is often considered a typical evolutionary sequence. Parallel evolution can be seen between *Coccochloris* and *Gloeocystis, Agmenellum* and *Prasiola, Entophysalis* and *Characium, Anacystis* and

Pleurococcus, Lyngbya and *Hormidium, Stigo-nema* and *Wittrockiella.*

As a group, the plants are extremely widely distributed under all sorts of conditions, frequently occurring in places where no other vegetation can exist, e.g. hot thermal springs. Their presence in great abundance in the plankton often colours the water and is responsible for water blooms, which in some cases may result in the death offish. Members of the group also form a large constituent of the soil algae.

One odd site that has been recorded for these algae is in the centre of lake balls composed of the remains of leaves of *Potamogeton pectinatus* found in Lake Vrana, Yugoslavia. Symbiosis involving Cyanophyceae is of widespread occurrence. Apart from the very well known example of Lichens (an association of a fungus and either a Chlorophycean or Cyanophycean alga, e.g. *Anacystis, Scytonemd),* blue-green algae have been recorded in a variety of symbiotic associations, examples are: *Anabaena azollae* within the leaves of the water fern *Azollafiliculoides*\' *Nostoc sphaericum* within cells of the fungus *Geosiphon pyriforme; Calothrix chapmanii* within the cells of the green alga *Enteromorpha.* One of the most interesting examples of symbiosis involving blue-green algae is the case of the 'cyanellae' or 'syn-cyanomes'.

These are blue-green algae living within colourless algae or rhizopods. The symbiosis appears to be geologically ancient and unlike other symbiotic Cyanophyceae. The algae have undergone such modification and reduction that some workers have suggested that in certain cases (e.g. *Glaucocystis)* the association is so inseparable that it may be considered as a single entity, and consequently they have suggested the creation of a phylum Glaucophyta to accommodate them. Another possibility is to regard them as only peculiar plastids. Two of the best known examples are *Glaucocystis nostochinearwn* (the host is a colourless Chlorophyceae-like organism with vestigial flagella) and *Cyanophora paradoxa.*

In this latter case there are two blue-green inclusions, which have been named *Cyanocyta korschikoffiana* (Fam. Cyanocytaceae, order Chroococcales), inside a colourless Cryptomonad host, *Cyanophora.* In the case *of Glaucocystis,* the name *Skujapelta nuda* (Fam. Skujapeltaceae, order:

Chroococcales) has been applied to the symbiont, and *Glaucocystis* retained for the host. The symbionts differ from free-living forms in lacking the typical, double-layered cell wall and they also possess a large centrally-located body which may represent a primitive sort of nuclear material. Another of these peculiar forms is the colourless parasite, *Sarcinastrum urosporae* found in *Urospora* and *Ulothrix*.

The Cyanophyceae, besides being primitive, are very ancient and some of the earliest plant fossils known belong to the class. It would seem that very little evolution has taken place within the group since they first appeared. Their age and constancy of form, together with their modes of reproduction, are also responsible for the wide distribution of different species, so that there is little or no endemism in the Cyanophycean flora of any region.

Taxonomically the class has been variously treated. Originally it was divided into two orders and in more recent years it* has been customary to recognize five orders. Much of the problem in this respect arises from environmentally induced morphological variation. Recent works on *Nostoc* and *Lyngbya* have shown that gross morphology is especially dependent on environmentculture condition. Drouet has shown that some 54 different species are in fact all variations (ecophenes) of *Schizothrix calcicola*. Some authorities, however, have suggested that there should be only one order, the Chroococcales, and eight families.

The five orders recognized in many taxonomic works are first, the Chroococcales, which comprises unicellular, free-living forms, living singly or in colonies; second, the Chamaesiphonales, which is a small order consisting of attached unicells occurring singly or in groups and reproducing by means of endospores, which do not occur in the first order. The Pleurocapsales is another very small order with few species that consist of cells that divide in one or two planes to give a crust or cushion.

The great majority of the Cyanophyceae are placed in the Nostocales and these are the simple or branched filamentous forms that occur free or aggregated in a common sheath. In some genera the base and apex of the filaments differ, e.g. *Rivularia, Calothrix,* and the possession of heterocysts is a feature of many. Reproduction is commonly by means of hormogones but some genera also produce spores. The fifth order is the Stigonematales in which the species form branched threads and there are examples of heterotrichy. Branching is true branching and the threads may be multiseriate. Both hormogones and heterocysts are known.

CHROCCOCCACEAE: ANACYSTIS

The representatives of this family consist of uni- or multicellular microscopic plants or macroscopic plant colonies. The individual cells are spherical, ovoid, discoid, cylindrical or pyriform in shape, and if colonial can be arranged regularly or irregularly within a gelatinous envelope. In some species of *Anacystis* the colonies are free-floating with a mass of spherical cells: in other, mostly terrestrial, species the cells are single or else united into spherical or flattened colonies each with a small number of cells.

The free-floating colonies of *Agmenellum* form regular plates one cell in thickness which, with increasing age and size, can become curved or twisted. There appears to be considerable variation in the degree of compactness with which the cells are arranged.

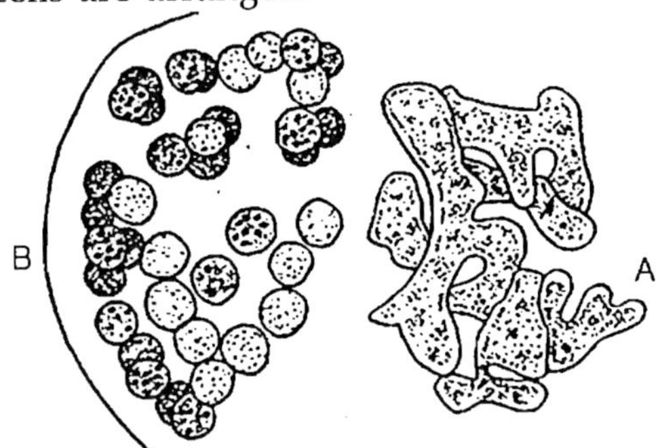

Fig. Anacystis Cyanea.

Reproduction of colonies is by fragmentation, but multiplication of the cells is by binary fission in two *(Agmenellum)* or three planes *(Anacystis)*. In the genus *Johannesbaptista*, which has a linear series of lenticular cells in an elongate gelatinous matrix, division is in one plane only. *Anacystis cyanea* is a very common water bloom alga and certain strains produce a highly toxic substance. The taxonomy of the family is extremely difficult and one can either recognize a large number of genera and species or only a few.

The latter course, based on Drouet and Daily, has been adopted here. The common *Anacystis nidulans,* of physiological use has recently been transferred to a new genus *Lauterbornia nidulans.*

CHAMAESIPHONACEAE: ENTOPHYSALIS

The cells of the sole genus in the family are attached or epiphytic, solitary or arranged in dense clusters on wood, rocks or fresh water or marine plants. The cells are more or less rigid, vary much in shape and are attached at the base by down-growing filaments.

In a cushion cell, division is usually horizontal and gives rise to a pseudo-parenchymatous prostrate system, but the upper daughter cells can divide by perpendicular walls, the resulting arrangement of cells being radial. The sheath is thin and hyaline and ultimately opens at the apex. Reproduction is by means of endospores which are abstricted successively from the apex of the surface cells by transverse division. There has been much confusion about this order, but Drouet and Daily consider that species previously placed in different genera, e.g. *Chamaesiphon, Dermocarpa, Xenococcus, Pleurocapsa,* simply represent growth stages of one and the same species. The few species recognized are now placed in the single genus *Ento-physalis.*

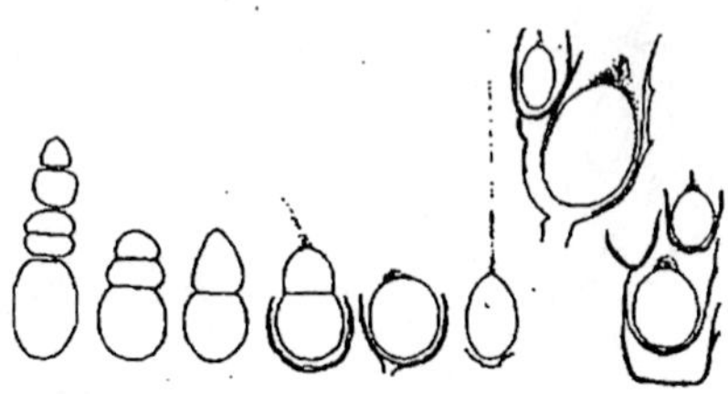

Fig. Clastidium Setigerum.

CLASTIDIACEAE

This family is not well known, the plants consisting of solitary, elongate, epiphytic unicells within a sheath which is thickened at the base to form an attachment pad. There are only two genera, and three species.

STIGONEMATACEAE: STIGONEMA

The members of this family, which includes *Mastigocladus*, form branched threads, sometimes with a prostrate and erect portion. The entire plants are floccose, felt-like, cushion shaped or spherical. Branches are formed by cell division parallel with the axis and the daughter cell pushes through the sheath at right angles and then produces its own sheath.

In Stigonema the threads consist of a single row or multiple rows of cells with definite apical growth, enclosed within a gelatinous sheath that varies in thickness and colour. Shortly behind the apical cell one of the products divides twice to give a cruciform group. One of these cells gradually moves into the centre and by further division a central cell surrounded by four pericentrals is produced.

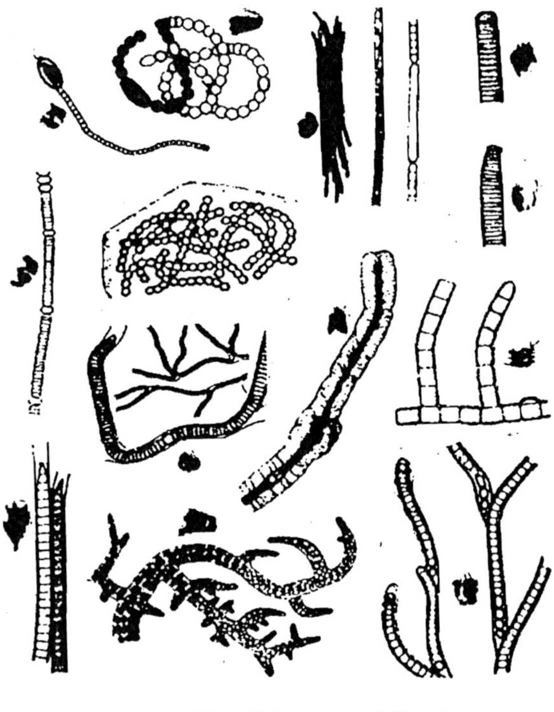

Fig. Stigonema Minutum.

Most pericentrals divide radially so that ultimately successive tiers of 7-8 pericentral cells are formed and these are arranged approximately longitudinally. In this genus the cells are connected by protoplasmic strands but the exact nature of these requires further study. In young plants the sheath is continuous at the tip, but when growth is rapid or when hormogones are produced and extruded the end bursts.

The heterocysts in this genus are commonly lateral, being formed by tangential division of a cell. The filaments are often aggregated to form a macroscopic yellow-green gelatinous mass. The species are most common where there is continual dripping water.

NOSTOCACEAE: NOSTOC

The gelatinous thallus is solid or hollow, floating, attached or terrestrial and varies much in size and shape. The spheres usually break open when mature and give rise to a flat expanse. There is a dense limiting layer containing numerous intertwined and contorted moniliform filaments with individual hyaline or coloured sheaths, though the sheath may sometimes be absent or very indistinct.

Heterocysts are terminal or intercalary and occur singly or in series. Reproduction is by means x>f hormogones or spores, the latter arising midway between the heterocysts and developing centrifugally.

The closely related genus *Anabaena* differs from *Nostoc* in that no firm colony is formed and it lacks a punctiform stage, whilst in *Wollea* the saccate plants contain trichomes arranged in a parallel manner. A feature of the genus *Cylindrospermum* is the large spore that develops next to the heterocyst. Some species are symbiotic, whilst species of both *Nostoc* and *Anabaena* are capable of fixing atmospheric nitrogen.

OSCILLATORIACEAE: OSCILLATORIA

This is the largest family of the class. It differs from the other families in that the members do not form spores or heterocysts.

The trichomes are uniseriate and occur singly or many in the enveloping sheath. Branching when present is of the false kind. All cells are capable of division and reproduction is by hormogones. In some cases it is very difficult to discern the sheath, e.g. *Spirulina,* some species of *Oscillatoria.* The filaments occur singly or else they form tangled masses. In the genus *Lyngbya* the sheath is often thick and coloured. Trichomes of *Oscillatoria* can exhibit a gliding motion based on travelling waves on the cell surface produced by a fibrillan array.

3
Chlorophyta

PRASINOPHYCEAE

This class was proposed only recently, mainly as a result of discoveries using electron-microscopy. Much of the work in this field has been carried out by Manton and Parke and their co-workers. The organisms are flagellates containing chlorophylls *a* and *b* and green algal xanthophylls, and are now known to be characterized by one or two layers of plate-like scales on the flagella and in some cases (e.g. *Monomastix, Micromonas)* on the cells as well.

There may be a single flagellum *(Micromonas squamatd)*, two equal flagella *(Nephroselmis gilvd)* or four flagella *(Prasinocladus)*. Hairs are also present on the flagella; these are thick and tapering, often curved at the tips, and usually with internal striations. In the genus *Heteromastix* the origin of the hairs has been clearly traced to vesicles derived from golgi bodies.

In *Heteromastix, Halosphaera* and *Micromonas squamata* the scales also originate from the vesicles and are subsequently deposited on the outside. Rounded plate scales form the inner of two scale layers in *Prasinocladus, Pyramimonas, Halosphaera* and *Heteromastix*. The outer scales may be small rods *(Prasinocladus)*, stellate *(Heteromastix)* or hexagonal *(Halosphaera)*. In *Nephroselmis* there is only one layer of scales. In *Tetraselmis* a product of photosynthesis is mannitol, normally produced in the Phaeophyta, and if this is confirmed for other members, then their segregation will be further justified on physiological grounds.

Asexual reproduction is by fission, usually into two daughter cells. In *Micromonas squamata* a second flagellum develops before fission, whereas in *Nephroselmis* and *Heteromastix* each daughter cell receives one parent flagellum and one newly grown one.

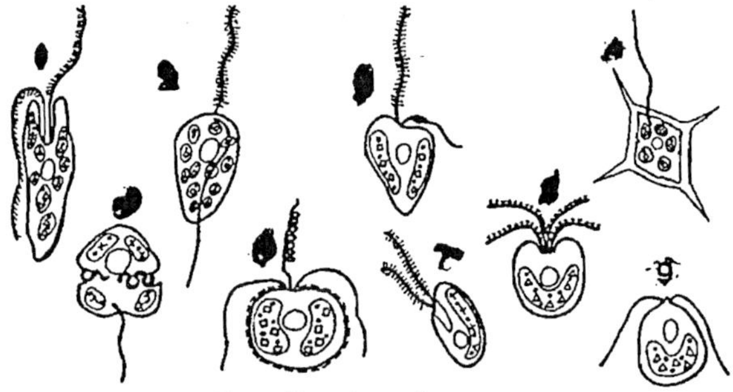

Fig. Prasinophyceae.

The exact confines of this class have not yet been finally established. Some authorities retain members of the Prasinophyceae *(sensu lato)* as three sub-orders of the Volvocales within the Chlorophyceae. Some would include *Pedinomonas* even though there are no scales on the flagella. Other workers place this genus and *Micromonas pusilla,* where there is a long hair point to the flagella and no scales, in a separate family, the Loxophyceae. The class has been tentatively divided into two orders, the *Pyramimonadales* and the *Halosphaerales.* Two genera that have been studied in some detail are Prasinocladus and Halosphaera.

CHLORODENDRACEAE: PRASINOCLADUS

Species of this genus are found in marine aquaria and on rock surfaces near high-water mark. The motile phase is a quadriflagellate swarmer with a laciniate parietal plastid, whose finger-like projections point anteriorly and posteriorly. The four flagella arise from the bottom of a depression at the anterior end. There is a distinct starch sheath surrounding the pyrenoid and also a 2-layered red eye spot which is more pronounced in summer swarmers than in winter ones.

The cells are covered by a periplast and when the organism becomes non-motile this undergoes plastic deformation to form the stalk of the sedentary individual. Because it contains no cellulose, this type of 'wall' may be termed a theca. The theca is composed of three layers, a central homogeneous one and an outer and inner fibrillar layer. Motile bodies are not common. When a cell becomes motile it starts to move out of the parent theca but becomes wedged in the neck, and then a new theca is formed. In this manner a series of empty cups are built up, and the plastic deformation of these empty thecae gives the jointed appearance of a hollow stalk.

Division of the non-motile cells takes place by longitudinal fission and this can result in branch formation; sometimes one daughter cell does not emerge but is left behind in the middle of a series of empty thecae. On the few occasions when a swarmer is liberated the species is perpetuated, but at present the conditions under which liberation occurs are not known. Cyst formation may also occur. The well known symbiont of the platyhelminth *Convoluta roscoffensis* has recently been reassigned to this order as *Platymonas convolutae.*

HALOSPHAERACEAE: HALOSPHAERA

The large, free-floating spherical cells possess one nucleus which is suspended either in the central vacuole or in the parietal cytoplasm where it is associated with numerous discoid chloroplasts. A new membrane can be formed internally and then the old one ruptures, but if the outer membrane persists the cells appear to have a multi-layered sheath.

Aplanospores and resting cysts have been recorded, but reproduction is primarily by means of quadriflagellate swarmers. These pear-shaped swarmers very closely resemble known members of the genus *Pyramimonas* both in respect of shape and morphology. There are anterior body lobes which can be outwardly or inwardly curved or which remain vertical. Around the body there are muciferous organelles or trichocysts which can eject their contents as rods, threads or balloons when treated with vital stains. Two or four pyrenoids with starch sheaths are also present.

The most distinct structure, however, is a large spheroidal basal reservoir with a canal to the exterior through the frontal lobes. There are two layers of scales but the scales of flagella and body differ in pattern. The outer flagella scales are round or hexagonal plates strictly aligned in nine rows. The second layer consists of much smaller rounded scales, each with a circular groove and a central hub, arranged in a spiral fashion.

The cell body scales are more varied but there is a principal diversity between flat lace-like plates and rectangular or circular scales with deeply depressed centres.

Cells of fertile thalli contain much more carotene than those of vegetative thalli, and a large proportion is y-carotene of which only a trace occurs normally. Other colouring matter may also be present, e.g. tannins in some of the Conjugales. Formation of secondary carotenoids is very common in the Chroococcales, and other isolated genera (e.g. *Acetabularid).*

The cells are commonly surrounded by a two-layered wall, the inner, which is often lamellate, being of cellulose (xylan in certain Siphonales), and the outer of pectin. In some forms the outer surface of this pectin sheath is dissolved as fast as it is formed on the inner side. Some of the very primitive forms lack a rigid cell wall.

In a few species the outer layer of pectose becomes impregnated with an insoluble substance while in others the pectin layer gradually increases in thickness, and in at least three orders (Siphonales, Dasycladales, Siphonocladales) lime may be deposited on the walls. In *Oedogonium* and *Cladophora* there is said to be a layer of chitinous material on the very outside and a study of the sub-microscopic structure of filamentous species has shown that there may be more than one kind of cellulose in the walls.

The chloroplasts normally contain pyrenoids surrounded by a sheath of starch. The pyrenoids contain fibrillar particles, but the existence of proteins is not substantiated by recent work though RNA may be present. The number of pyrenoids in each cell is frequently a constant specific character and can be used for taxonomic purposes.

Very few families of Chlorophyceae lack pyrenoids and those that do are the more highly evolved. In *Derbesia* the pyrenoids disappear when the alga is grown in diffuse light, and in several genera they may disappear during zoospore formation. In such cases they must arise again *de novo,* but more commonly they are perpetuated by simple division at the time of chloroplast division. In those algae with pyrenoids the production of starch granules is associated with their presence.

Each cell is usually uninucleate, but in certain orders a

multinucleate condition is found, and in the Siphonales cross walls are only laid down at reproduction so that the vegetative plant is non-septate and multinucleate. The chromosomes are usually small, short and few in number. Where there is only one chloroplast it divides at cell division into two daughter chloroplasts. In many species there is a distinct diurnal periodicity in nuclear and cell division.

The flagella of the motile cells are composed of an axial, cytoplasmic filament or axoneme surrounded, except at the very apex, by a sheath which is probably able to contract. In *Chlamydomonas* and the Volvocaceae the electron microscope has revealed that the axoneme consists of 9 separate fibrils arranged in a ring around 2 central fibrils. This 9 + 2 arrangement is general throughout the algae, except in the haptonema of the Haptophyceae.

In the motile forms the flagella are associated with other structures which are collectively known as the neuromotor apparatus. At the base of each flagellum is a granular blepharoplast. These are connected by a transverse fibre, or paradesmose, and this, or one of the blepharóplasts, is also connected to the intranuclear centrosome by a thin strand called the rhizo-plast. The motile cells often contain a red-eyespot. In *Platydorina* the eye-spot comprises an assemblage of carotenoid-containing granules situated between the lamellae of the chloroplast.

In other algae the carotenoid granules may have an association with the flagella with (Chrysophyceae) or without plastid association (Euglenophyta), or with flagella and a lamellar body *(Glenodinium foliaceum)* or associated with a retinoid-like structure (Warnowiaceae in Dinoflagellates).

The exact function of the eyespot has not been determined, although it is possible that it acts as a shading device for the parabasal granule. Vegetative reproduction takes place through fragmentation and ordinary cell division, though in the Chlorococcales and Siphonales vegetative cell division is unknown. Asexual reproduction is by means of biflagellate or quadriflagellate zoospores, except in the Oedogoniales where there is a ring of cilia at the anterior end.

The zoospores are often formed during the night and are then liberated in the morning when they may remain motile for as much as three days or for as short a time as three minutes. Their production can sometimes be induced artificially by altering the environmental conditions, e.g. by removing the plant from flowing to still water *(Ulothrix, Oedogoniuni)*, by changing the illumination, or by removing the plant from water for twenty-four hours *(Ulva, Enteromorphd)*. Each individual cell may produce one or more zoospores, the number varying with the different species.

Liberation depends on:

1. Lateral or terminal pores,
2. Gelatinization of the entire wall,
3. The wall dividing into two equal or unequal portions.

In some species non-motile zoospores called aplanospores are formed, but if these then secrete a thick wall they become known as hypnospores. Aplanospores which have the same shape as the parent cell are termed autospores. All these spores develop a new periplast or membrane and therefore differ from a purely resting vegetative cell or akinete.

Sexual reproduction is represented in all the orders and often there is a complete range from isogamy to oogamy, the egg sometimes being retained and fertilized on the parent thallus in the oogamous forms (e.g. *Volvox, Oedogonium, Coieochaete*, Charales). The isogamous forms are normally dioecious, the two sexes being termed 4- and -, because as plants and gametes are alike morphologically they are only distinguished by the fact that the gametes can normally only fuse with gametes derived from another plant.

In some cases *(Viva, Enteromorpha, Chlamydomonas)* relative sexuality has been reported, weak + or - strains fusing with strong 4- or - strains respectively. Segregation into + and - strains occurs during meiosis, a phenomenon which in many species takes place at the first division of the fertilized egg or zygote. In the Dasycladales the gametes are produced in special cysts that are liberated from the parent thallus. In the case of oogomous algae, the gametangia (antheridia, the male, and oogonia, the female) are normally of distinctive shape. In some genera gametes are capable of developing into new plants by parthenogenesis.

The occurrence of sexual reproduction can be induced in culture by an abundance or deficiency of food material or by intense insolation. Interspecific hybrids have been recorded in *Spirogyra, Ulothrix, Stigeoclonium, Draparnaldia,* and *Chlamydomonas.* Another striking fact is that characters which may develop in some species under the influence of the external environment are normally found 'fixed' in others. This indicates the plasticity of many members in the class and the phenomenon may also be of importance phylogenetically.

As a group the Chlorophyceae are widespread, occurring in all types of habitat: marine, freshwater, soil and subaerial. Only about 13 per cent are marine and are mainly represented by the Ulvaceae, Siphonocladales, Dasycladales, Derbesiales and Siphonales. Many of the marine species have a definite geographical distribution whereas most of the freshwater and soil algae are cosmopolitan. A few genera, e.g. *Entocladia (Endodermd), Chloro-chytrium, Cephaleuros, Phyllosiphon,* are parasitic species, whilst species of other genera participate in symbiotic associations, e.g. *Zoochhrella*

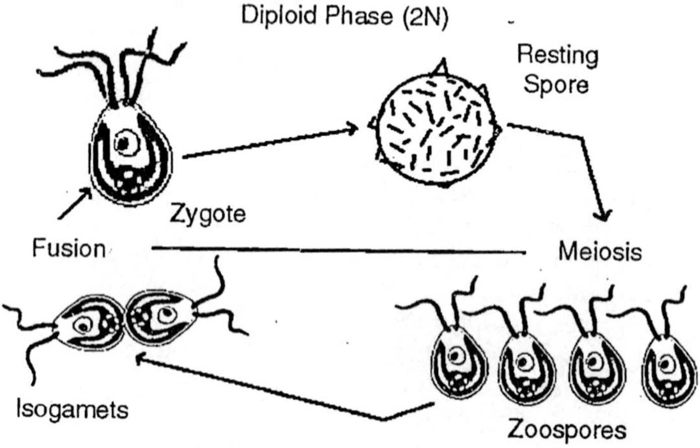

Fig. Chlamydomonas

CHLAMYDOMONADACEAE: CHLAMYDOMONAS

The chlamydomonad type of cell characteristically possesses a single chloroplast, a red eye-spot, one pyrenoid, two contractile vacuoles and two flagella, and is often strongly attracted to light (positively phototactic). Variations in the structure of the cell occur throughout the genus which contains about 325 species. Some species lack pyrenoids some have one *(C. stellatd)*, or two (C. *de baryand)*, or several (C. *giganted)*.

The chloroplast is usually basin-shaped but may be reticulate (C. *reticulata)*, or axile and stellate (C. *eradians)*, or it may be situated laterally *(C. parietarid)*. It has been said that under cultural conditions many of the characteristic features can be modified, and that therefore some of the species are not strictly species but are simply phases in the life cycles of other species, or ecophenes of the same species.

The motile cells are spherical, ellipsoid, ovoid or pyriform in shape with a thin membrane which sometimes has an outer layer of mucilage. The term naked is often used to describe cells (usually flagellates) which lack a cell wall (e.g. cellulose, pectin). The term is relative as so-called naked cells (e.g. *Chlamydomonas)* often possess an outer membrane, which surrounds the whole cell, and also possess a thin protoplasmic membrane or periplast. In this text the term naked describes cells that lack the cellulose-type wall characteristic of non-flagellates without denoting any particular type of periplastic membrane. The two flagella are situated anteriorly and either project through one aperture in the membrane or else through two separate canals.

In either case there are two basal granules or blepharoplasts at the point of origin of the flagella. Electron microscope studies have shown that every cell contains mitochondria, dictyosomes, a double cell membrane, endoplasmic reticulum and fenestrated nuclear membrane. Near the dictyosomes there are also large irregular vesicles whose function is unknown.

During asexual reproduction the vegetative cell comes to rest and divides into two or four, more rarely eight or sixteen, zoospores. All divisions at zoospore formation are normally longitudinal. The biflagellate zoospores escape through gelatinization or rupture of the cell membrane, but if this does not occur the colony then passes into a palmelloid state, which is usually of brief duration; in C. *kleinii* however, it is the dominant phase in the life of the organism. *C. kleinii* may thus be regarded as forming a transition to the condition found in *Tetraspora.*

In sexual reproduction 8, 16 or 32 biflagellate gametes

are formed in each cell. These are normally either + or - in character and fusion takes place between the strains. In C. *agametos* an anticopulatory hormone is produced and so there is no fusion; in C. *monoica there* is physiological anisogamy as the naked contents of one gamete (male) pass into the envelope of the other; in C. *braunii* the anisogamy is more distinct, the female cell producing two or four macrogametes and the male cell 8 or 16 microgametes.

In C. *suboogama* each cell gives rise to three macrogametes and an antheridial cell that produces four antherozoids; this is a most interesting condition, because sex segregation must take place at the first division of the mother cell nucleus. In C. *coccifera* there is oogamy, with the female cell producing one macro-gamete enclosed in a wall, while the male cell produces 16 or 32 spherical microgametes. In a related genus, *Chlorogonium oogamum,* one naked ovum is produced and numerous elongate antherozoids. The genus *Chlamydomonas* appears to be polyploid, with eight as the basic haploid chromosome number.

The zygote is at first quadriflagellate but it soon loses the flagella and forms a smooth or spiny wall. It subsequently enlarges in most species and on germination generally gives rise to four swarmers. Meiosis occurs during this segmentation and the normal vegetative cells are therefore haploid. In C. *pertusa* and C. *botryoides,* however, the zygote may remain motile for as long as ten days.

In C. *variabilis* the persistent 4-ciliate zygote has for long been known as *Carteria ovata* but it has now been demonstrated that the two organisms represent the haploid and diploid phases respectively of one species. Whether these examples represent a true alternation of generations or whether they must be regarded as possessing a special type of zygote will be considered later. In most species the zygote requires a maturation period, which in C. *reinhardii* is 6 days at 25°C. The life history, which probably holds for the majority of the other species. The species is very easy to grow in culture and the four products of zygote germination readily separate on an agar medium.

Many mutants of C. *reinhardii* have appeared during investigations and it has been possible to work out a number of linkage groups of different genes. Many of the mutant strains are of such a nature that it will be possible to attack fundamental problems in the biochemistry as well as the genetics of these organisms.

The genus is widespread, the various species occurring principally in small bodies of water, in sewage oxidation lakes and in the soil.

Within the family Chlamydomonadaceae there is a considerable variety of forms. Among those which lack a true cell wall is the biflagellate *Dunaliella,* one species of which, *D. salina,* is very common in salt ponds. In *Phacotus* the chlamydomonad cell is surrounded by a special outsize hardened envelope which is flattened in side view. *Polytoma* is a colourless member of the family but nevertheless produces starch as its food reserve.

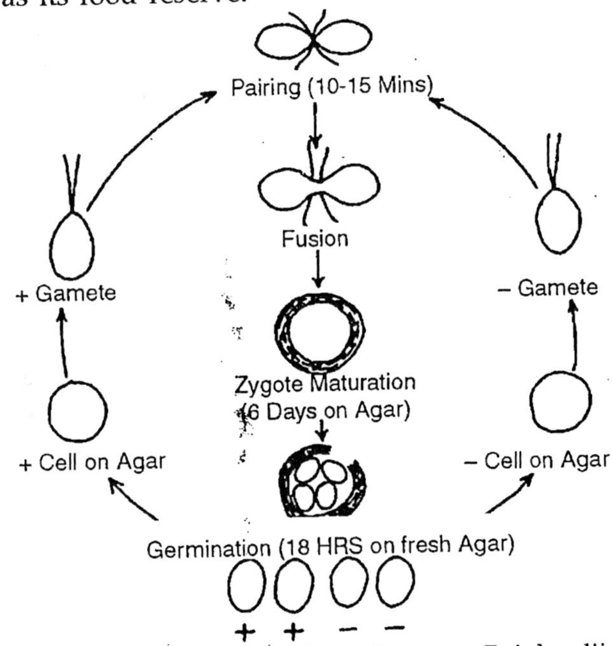

Fig. Life History of Chlamydomonas Reinhardii.

HAEMATOCOCCACEAE: HAEMATOCOCCUS

A characteristic of this genus is the area forming the inner wall between the protoplast, which is very similar to that of

Chlamydomonas, and the firm, outer cell wall; this region is filled by a watery jelly and crossing it are cyto-plasmic threads which pass from the central protoplast to the cell wall.

These are numerous and very fine in *H. pluvialis,* but are fewer and thicker in the other species. The flagella are unusual in possessing a thin tomentum of fine hairs. Contractile vacuoles are not always present and if present they are inconspicuous. The plastid varies in the different species and the number of pyrenoids ranges from one in *H. capensis* to many in *H. pluvialis.*

In asexual reproduction the zoospores may pass through a non-motile aplanospore stage before maturing, and this has developed to such an extent in *H. pluvialis* that it forms a major feature of the life history. In all species except *H. pluvialis* small biciliate gametes can be formed, the number ranging from 32 in H. *buetschlii* to 256 in *H. capensis.* In H. *pluvialis* gametes are produced only rarely and never from motile cells, as in the other species, but only from resting cells.

VOLVOCACEAE

So far the genera considered have comprised only unicells, but a whole series is known in which an increasing number of chlamydomonad cells are combined into colonial forms. The simplest type is probably represented by *Pascheriella* where three or four chlamydomonad cells are attached to each other laterally. The colonies are usually spherical in shape, but in the genus *Gonium* four, eight, or sixteen cells all lying in one plane form a flat quadrangular plate.

In the genus *Pandorina* 4, 8, 16 or 32 cells are closely aggregated in a common spherical envelope, whilst in *Vohulina* the 16 cells are arranged in four tiers.

This process has gone further in the genus *Eudorina* where the colonies are spherical or ellipsoidal with the posterior end often marked by mamillate projections. The colonies contain sixteen, thirty-two (commonly) or sixty-four biflagellate cells, which are not closely packed and are sometimes arranged in transverse rows. In *E. illinoiensis* the four anterior cells are much smaller and cannot reproduce.

This marks the first differentiation of this series of organisms into a plant soma of purely vegetative cells which die once the colony has reproduced. Further differentiation occurs in *Pleodorina* where the somatic cells form J to i of the 64- or 28-celled colonies.

A genus comparable to *Pleodorina* is *Astrephomene* which is found in S. Africa, Australia and the U.S.A. The colony consists of 64 cells, the four anterior cells forming a 'rudder'. In these colonial forms asexual reproduction involves the formation of daughter coenobia from all the cells except somatic cells.

Sexual reproduction in *Pandorina* shows a tendency to anisogamy, but in *Eudorina* it borders between anisogamy and oogamy, the species being either monoecious or dioecious: in the former case the anterior cells give rise to the antherozoids and the posterior form the ova. When the zygote germinates it gives rise to a single motile swarmer and what are probably two or three degenerate swarmers. The motile swarmer eventually comes to rest and divides mitotically to give a new coenobium.

VOLVOX

The genus represents the ultimate development that has been reached along this line of spherical colonies or coenobia, each of which behaves as a well-organized unit. Every colony is in the form of a sphere with 500-60,000 biflagellate cells set around the periphery, the flagella emerging through canals. The actual number of cells in a colony varies from species to species. The interior of the colony is mucilaginous, as in *V. aureus,* or else merely contains water, as in *V. globator,* while the entire collection of generally ovoid cells is bounded by a firm mucilaginous wall. The individual cells, each containing two to five contractile vacuoles, are surrounded in some species by distinct individual sheaths.

In some species, too, the cells are united by delicate cytoplasmic threads, or plasmodesmae, and the base of each thread appears to be associated with a mitochondrion. In *V. globator* and *V. rousseletii* the cells are 'sphaerelloid' in nature, each individual cell being enclosed in a separate envelope, while in *V. aureus* a number of individual chlamydomonad cells are enclosed in wedge-shaped prisms. It has therefore been suggested that the volvocine colony has arisen at least twice in the course of evolution, once from a *Haematococcus* and once from a *Chlamydomonas* ancestry.

The majority of the cells, including all those in the anterior quarter, are wholly somatic, and only a few are able to give rise to daughter colonies or gametes. As the plant matures some of the posterior cells enlarge and lose their flagella. The enlarged gonidium then divides longitudinally a number of times and a small hollow sphere is produced with a pore (phialopore) towards the outer edge. These coenobia, which hang down into the parent cavity, then invert, the process commencing opposite the phialopore.

After inversion flagella are formed and the colonies may be liberated into the parental cavity or they may escape to the exterior. If retained they remain until the parent tears open, in some species *(V. aureus)* at the adult phialopore, in other species *(V. globator)* at any place. In *V. africana* it is possible to see as many as four generations in the one original parent colony, because the first parent takes a long time to break down.

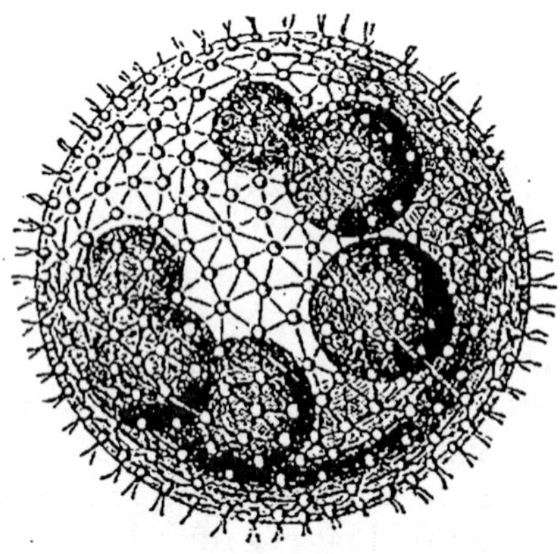

Fig. Volvox

In sexual reproduction the plants are either monoecious *(V. globator)* or dioecious *(V. aureus)*. Cells which give rise to antherozoids divide to give bowl-shaped or globose colonies containing 16, 32, 64, 128, 256 or 512 spindle-shaped, biflagellate antherozoids. Starr reports that in eight strains of *Volvox* the liberation of a male proteinaceous hormone induces egg formation.

Cells developing into ova enlarge considerably and loose their flagella but do not undergo division. Both bowls and spheres undergo inversion before maturation. The fertilization mechanism is not known for certain, but in *V. aureus* the antherozoids penetrate the female colony and then enter the ovum from the inner side.

Zygotes do not germinate for a considerable time after liberation. Meiosis occurs at germination but the oospore only produces a single swarmer that is liberated into a vesicle formed by extrusion of the inner oospore wall (endospore). This grows into a 'juvenile' plant of about 128 to 500 cells, which finally inverts. It subsequently reproduces by daughter colonies, each successive generation having an increased number of cells until a fully developed colony is produced.

One of the features of the larger saccate Volvocales (excepting *Astrepho-mene)* is the inversions that occur at different stages of the life-cycle, and it is difficult to see why they occur or what the conditions were under which they first developed. It may have something to do with the fact that the new cells are formed with the flagella and eye-spot facing the interior, but then the problem arises as to how the individual cells came to be arranged thus.

The nuclear condition of the Volvocaceae presents some interesting features. *Gonium,* which would seem generally to be a primitive member of the family, possesses the largest number of chromosomes. In *Volvox* the species regarded as primitive have more chromosomes than species regarded as advanced ($n = 5$). The facts may necessitate a re-interpretation of development within the family, *Gonium* being regarded as reduced rather than primitive, and similarly with the species of *Volvox.*

TETRASPORACEAE: TETRASPORA

Members of this and allied families are placed by some workers in a separate order, the Tetrasporales. They are characterized by non-motile vegetative cells that may temporarily become motile. In so far as the cells are chlamydomonad-like, there seems no justification for the segregation. The species of *Tetraspora* form expanded or tubular, convoluted, light green macroscopic colonies. The colony in these palmelloid forms differs from the organized coenobium of the Volvocaceae because it is merely an unorganized loose assembly of cells.

The palmelloid condition has in this and allied genera become the dominant phase, while the motile condition only occupies a brief period in the life cycle. The colonies are most abundant in the spring when they are attached at first, although later they become free-floating. The spherical to ellipsoidal cells are embedded in the mucilage and are frequently arranged in groups of two or four, each group being enclosed in a separate envelope. Two pseudo-cilia proceed from each cell to the surface of the main colonial envelope, each thread being surrounded by a sheath of denser mucilage.

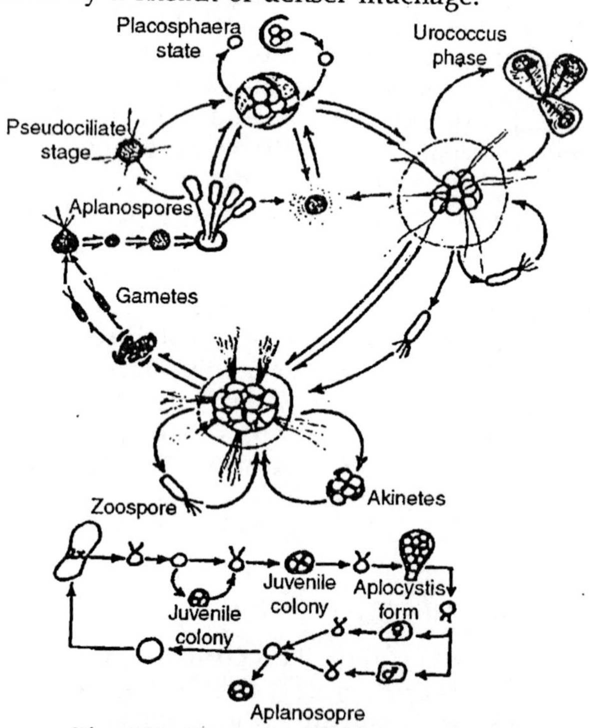

Fig. Life History of Schizochlamys.

Vegetative reproduction takes place by fragmentation of the parent colony, and asexual reproduction is secured by means of biflagellate swarmers that may develop into a new colony or into a thick-walled resting spore that later gives rise to a new colony.

The gametophytic colonies are monoecious or dioecious, depending on the species, and give rise to biflagellate isogametes. The zygote gives rise to four to eight aplanospores that are said to grow into new colonies. Work on *T. gelatinosa* suggests that the aplanospores are not always produced and that the zygote, after a resting period, can give rise directly to a new diploid plant which gives rise to asexual swarmers. These develop into juvenile colonies that can perpetuate themselves by biflagellate swarmers.

Eventually swarmers are produced that give rise to an attached, pear-shaped colony. The swarmers from this colony give rise to the sexual plant so that in this species there is a definite alternation of morphological phases. It seems probable, though it has not been yet established, that meiosis may well take place at swarmer formation in the attached colonies.

A life cycle has been fully worked out for *Schizochlamys*. Here the adult can reproduce itself by akinetes, zoospores or gametes by means of motile isogametes, but oogamy has been recorded for a species of *Dictyosphaerium*.

CHLOROCOCCACEAE: CHLOROCOCCUM

Much confusion has existed over this genus, and many of the species formerly described are now known to be phases in the life cycles of species from other genera. Bold and Parker consider that there are 18 valid species which are not always readily distinguished on morphological grounds, in which case changes in colony shape, cell size, colour or colony texture under specific culture conditions may be useful.

The plants are represented by non-motile, spherical or ellipsoidal cells which vary in size, occurring singly or else forming a stratum on, or in the soil; *C. humicolum*, is a very common soil species. There is no eye spot or contractile vacuole; the chloroplast is hollow and parietal with one or more pyrenoids. The characteristic features of this genus are first, the hollow, parietal chloroplast; second, the presence of one or more pyrenoids; and third, biflagellate zoo-spores that retain their oval shape for some days after losing their motility.

The cell walls are two-layered with a thin inner layer and an outer gelatinous one which is sometimes lamellose and irregularly thickened. Old cells may be coloured with an excess of secondary carotenoids. The young cells with thin walls are uninucleate but the adult ones with thick walls are commonly multinucleate, and it is in this older condition that the protoplast divides and gives rise to 8, 16 or more biflagellate zoospores or isogametes which are liberated all together in a vesicle, usually in the early hours of the morning. After a short motile phase the flagella are withdrawn and a new vegetative phase commences.

Under dry conditions aplanospores are formed and when this occurs the parent cell gelatinises and a 'palmella' stage results, the cells of which subsequently give rise to 2 or 4 biflagellate gametes. In culture solutions of low nutrient concentration, reproduction takes place by means of zoospores, but in highly concentrated solutions these are replaced by aplanospores. The environment can therefore affect the mode of reproduction. The production of aplanospores suggests how the genus *Chlorella* may have arisen, while the multinucleate state indicates a transition to a coenocytic phase that is further developed in other genera.

In *Trebouxia*, the cells differ from those of *Chlorococcum* in lacking an irregularly thickened wall and the plastid is axile rather than parietal. Certain species of the genus *Codiolum*, have now been shown to be stages in the life history of *Spongomorpha* and *Acrosi-phonia*. *Codiolum kuckuckii* is restricted as an epiphyte to the apical region of the southern brown alga *Splachnidium*.

CHARACIACEAE: CHARACIUM

Each plant is a solitary unicell with a single parietal chloroplast, and motile only in the reproductive phase. The ellipsoidal cells occur singly, or in aggregates on submerged plants or living aquatic larvae and they are borne on a short stalk that emerges from a small basal disc. In most species the mature cells contain more than one nucleus.

Asexual reproduction is by means of biflagellate zoospores which are liberated through a terminal or lateral aperture. Sexual reproduction is generally isogamous but some species exhibit anisogamy. In *C. saccatum* the -f and - gametes are produced from separate plants and fuse to give a zygote which, on germination, divides to produce a number of zoospores.

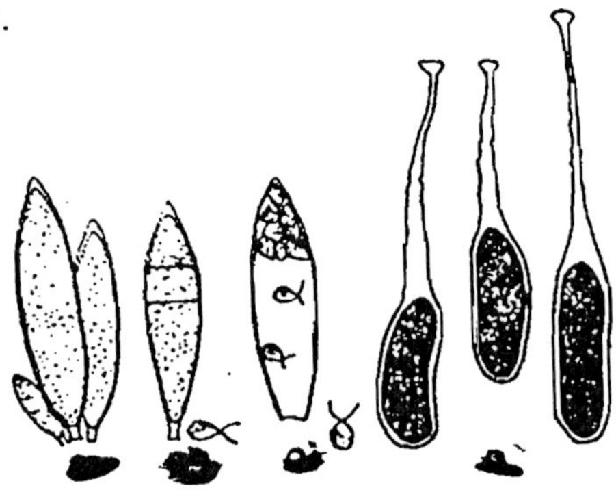

Fig. Characium Angustatum

ENDOSPHAERACEAE: CHLOROCHYTRIUM

The species grow endophytically in other plants such as mosses, angio-sperms and red algae. The cells are spherical or ellipsoidal, the walls varying in thickness. Reproduction is by means of biflagellate gametes that normally fuse to give a motile zygote. This eventually settles on the leaf of a host plant and secretes a wall. Tubular prolongations grow out from these bodies and enter the host, either by the stomata or between two epidermal cells.

Subsequently the end of the tube swells out into an ellipsoidal or lobed structure into which the contents of the swarmer pass. In *Chlorochytrium lemnae;* which attacks duckweed *(Lemnd)*, the endoparasitic cells sink in the old *Lemna* fronds to the bottom of the pond or stream and remain dormant until the next spring. There is evidence that reduction division takes place at gamete formation so that the vegetative phase is therefore diploid.

CHLORELLACEAE: CHLORELLA

The globular cells are non-motile, solitary or aggregated into groups, and usually lack pyrenoids. The species are not readily determined taxonomically, so cultures are often given strain numbers. Using morphological and physiological criteria some eight well-defined species can be recognized. They will grow in a wide range of media, including salt solution (0-2 M.). It seems that cells grown in the dark differ from those grown in the light. Quiescent dark cells are small and have a photosynthetic quotient (O_2/CO_2) of 1.

They give rise to dividing dark cells which in daylight give rise to 'light' cells that are larger, poorer in chloroplast material and less active photosynthetically.

During the course of development the amount of DNA increases. 'Light' cells give rise to a new generation of 'dark' cells in the dark and reproduction is by means of division into 2, 4, 8 or 16 auto-spores. Several species form a symbiotic association with lower animals and they are then known as zoochlorella. Because the species grow well in uni-algal culture they are often used in physiological investigations; they are also used in sewage purification.

COELASTRACEAE: SCENEDESMUS

The planktonic colonies are composed of four, eight, or, more rarely, sixteen cells attached to each other at one point by mucilage pads. This type of colony has probably originated from that in which cells are attached at one point indiscriminately, as in *Ankistrodesmus.* A large number of species are known to occur widely, and often in some abundance, in ponds and lakes.

Taxonomically the two end cells are highly important. They may differ in shape from the others and often have processes that are elaborations of the cell envelope (e.g. *S. quadricaudd);* these processes can probably be correlated with the planktonic mode of life. Other species (e.g. *S. acuminatus)* have tufts of bristles that perform the same function and are similar to those of *Pedia-strum.*

The cell wall of *S. quadricauda* exhibits a complex structure under the electron microscope. On the outside there is a pectic, hexagonal net layer which encloses a complex prop system in pectin. There is a middle layer and an innermost cellulose layer. Reproduction takes place when the contents of each cell in the colony divide to give daughter coenobia which are then liberated.

HYDRODICTYACEAE: HYDRODICTYON

The algae included in the Hydrodictyaceae occur as coenobia and represent a more advanced stage of the simple condition seen in *Ankistrodesmus* and *Scenedesmus*. Species of *Pediastrum* are common components of freshwater plankton, with up to 128 cells united into flat, disc-like coenobia.

The most advanced condition is found in *Hydrodictyon* of which only three species are known, the commonest, *H. reticulatum,* being cosmopolitan in distribution. It is a hollow, free-floating, cylindrical network closed at either end and up to 20 cm in length. Further increase in the size of colony is probably impossible for purely mechanical reasons.

The chloroplast is reticulate with numerous pyrenoids, though in the young uninucleate cells there is only a simple parietal chloroplast which later becomes spiral and then reticulate. *H. africanum* and *H. patenaeforme* develop into saucer-shaped nets, and in the former, the original young cylindrical cells later become spherical and up to 1 cm in diameter, then become detached and lie on the substratum.

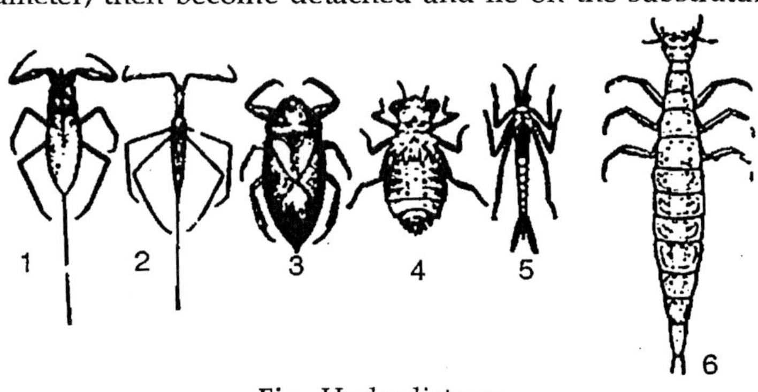

Fig. Hydrodictyon.

The latter species is composed of cells which may grow to be 4 cm long and 2 mm in diameter. Experiments have shown that for growth of *H. reticulatum* the actual growing period depends upon light intensity, e.g. it is 3 weeks at 600-700 Ix, 5£ weeks at 250 Ix. Under normal conditions growth rate is periodic, increasing by day and decreasing at night and altering the length of the light period merely alters the degree of growth in the time. Under conditions of continuous light the periodicity is much reduced, though this effect is more marked with old cells.

Asexual reproduction in *Pediastrum* and H. *reticulatum* is by means of numerous uninucleate biflagellate zoospores, which swarm in the parent cell about day-break and then come together to form a new coenobium which is subsequently liberated, further growth being brought about by elongation of the coenocytic cells.

Sometimes zoospores, instead of coming to rest to form a daughter net, settle down to form haploid hypnospores. These are not very common in *H. reticulatum* but are normal in *H. patanaeforme* and possibly also in *H. africanum,* where asexual reproduction of the coenocytes. In the first two species the hypnospore germinates to give a large, rather slow-moving biflagellate zoospore that settles down to form a polyhedron, whose contents later divide and after forming a pavement stage develop swarmers that are liberated into a vesicle to give a new 'germ net'. The duration of the 'germ net' in *H. reticuJatum* is short because each coenocyte rapidly produces a daughter net. This does not happen in *H. patanaeforme.*

Sexual reproduction is isogamous in H. *reticulatum* and *H. patanaeforme,* but anisogamous in *H. africanum.* In the first species the biflagellate gametes are smaller than the zoospores, but in the second species there is no distinction in size. Both species are generally monoecious. The planozygote arising from conjunction is quadriflagellate, and after a motile period it comes to rest and forms a diploid hypnospore. At germination the zygote generally enlarges and germinates meiotically to give 4, or more rarely 8, slow-moving biflagellate swarmers that later pass into the polyhedron stage.

Sexual reproduction in *Pediastrum* follows the same course. Gamete and zoospore reproduction can be induced in *H. reticulatum* by varying the external conditions artificially. If plants are cultured in weak maltose solution in bright light or in the dark and are then transferred to distilled water, zoospores will develop from the light plants and gametes from the dark ones.

Hydrodictyon is essentially a collection of a number of individual plants because it arises as the result of the fusion of a number of swarmers. *Volvox*, on the other hand, must be regarded as a single plant composed of many cells because it arises from a single zygote or gonidium.

PROTOSIPHONACEAE: PROTOSIPHON

There is probably but a single species, *P. botryoides*, but it exists in a number of different strains, one strain from the deserts of Egypt tolerating temperatures up to 91 ° C and salt concentrations up to 1 per cent. It commonly grows associated with a very similar alga, *Botrydium* in the damp mud at the edges of ponds. The green aerial portion is more or less spherical, grading into a colourless rhizoidal portion that may branch occasionally.

The multinucleate cell contains one parietal reticulate chloroplast with many pyrenoids. The shape and colour of the thallus can be modified by varying the external conditions; bright light and low moisture, causes an old thallus to turn brick red due to secondary caro-tenoid formation. During dry weather the contents of the vesicle encyst to give large aplanospores or 'coenocysts'. When conditions are once more favourable these cysts either grow directly into vegetative cells or else produce biflagellate gametes or zoospores.

Vegetative reproduction can take place by lateral buds which are cut off by cell walls. Flooding of plants on damp soil causes them to produce biflagellate swarmers which behave as isogametes though these swarmers can behave as zoospores. The zygote either germinates directly or may remain dormant for some time. The plant is probably haploid as there is evidence of meiosis at zygote germination. Morphologically it is of great interest as a possible source of origin for the Siphonales.

This order comprises filamentous and parenchymatous types which have clearly evolved from the filamentous form. There is little or no differentiation between base and apex in many of the forms though there is usually a special attachment cell or pad of cells. In most genera all the cells except those for attachment are capable of producing reproductive bodies, but in two genera from New Zealand *(Gemma, Lobatd)* there appears to be a considerable degree of sterilization. Within the genus *Enteromorpha* there is also evidence of the development of heterotrichy, or the condition in which the thallus possesses a prostrate adherent portion and an erect aerial portion.

Some workers have divided the genera into three orders: Ulotrichales, Ulvales and Schizogoniales or Prasiolales. Others have recognized only the first and last of these orders, or else have regarded them as forming two suborders, the Ulotrichineae and Prasiolineae.

Even when only one order is recognized some authorities accept only three families (Ulotrichaceae, Ulvaceae and Prasiolaceae) whereas others, including the present writers, suggest that there should be as many as eight families. The Ulotrichaceae and Microsporaceae contain only filamentous unbranched forms which can be separated from one another by chloroplast and cell wall structure.

The Cylindrocapsaceae are also unbranched but reproduction is oogamous as distinct from the isogamy or anisogamy in the first two families. The Sphaeropleaceae are also filamentous and oogamous but differ morphologically from the Cylindrocapsaceae. Some of the remaining four families consist of plants with a flattened or tubular parenchymatous thallus. Some members of the Monostromataceae are distinct because of the heteromorphic alternation and some of the remaining species *(Vivaria)* could either be included in the Ulvaceae or treated as a separate family.

The Prasiolaceae are distinguished by their thallus and plastid structure and the Capsosiphonaceae by thallus structure, habitat and reproduction. Until very recently the taxonomic treatment of the order was based upon morphology. It was then found that some species of *Monostroma* had a life history with a much reduced but persistent sporophyte stage, and so a separate family for *Monostroma* was proposed. Since then more attention has been given to life histories and Kornmann has suggested that the order should be restricted to species with an alternation of heteromorphic generations and to their derivative species.

Such a radical re-alignment brings together species of *Ulothrix, Vrospora, Gomontia* and *Monostroma*. Morphologically these genera differ widely, and there are also differences in their early stages of development. If therefore we can accept that alternation of heteromorphic generations may have arisen more than once in the course of evolution the traditional taxonomic treatment would seem preferable.

Despite the great interest of Kornmann's approach it would seem that it raises more problems than it solves. Another approach by Vinogradova that unites *Monostroma, Gomontia, Kornmannia* and *Enteromorpha (Blidingid) nana,* presents difficulties with respect to the last genus which is more closely related to *Enteromorpha.*

ULOTRICHACEAE: ULOTHRIX

The unbranched filaments are attached to the substrate by means of a modified basal cell which frequently lacks chlorophyll, but even though attached at first, the plants may later become free-floating. Under unfavourable conditions, such as nutrient deficiency, rhizoids may grow out from the cells or the filaments become branched. This suggests one way at least in which the branched habit may have evolved from the simple filament, and in this case it probably represents an attempt to increase the absorbing surface in order to counteract the deficiency of salts.

The cells of the different species vary considerably in size, shape and wall thickness. When the walls are thick they are often lamellate. The single chloroplast forms a circular band around the whole or most of the cell circumference and possesses two or more pyrenoids. When it does not encircle the cell completely the chloroplast margins are lobed. Vegetative reproduction can take place through fragmentation.

The life histories of the various species are not well known even though the genus has been recognized for a long time. There are two distinct groups of species: those in which there is an alternation of isomorphic generations and those in which only the haploid generation is known or in which there is a dwarf phase.

Swarmers are formed from all cells of the filament except the attachment cell, but they usually appear first in the apical cells and then in successive lower cells. They are liberated through a hole in the side of the cell into a delicate vesicle which subsequently bursts to free the swarmers. Asexual reproduction is normally by means of quadriflagellate zoospores, though in one species they are biflagellate. In the most recent revision of the genus Mattox and Bold transfer species with biflagellate zoospores to the genus *Hormidium.*

The numbers of zoospores produced per cell depend on the width of the filament: wide cells give 2, 4, 8, 16 and 32 zoospores per cell, narrow cells 1, 2 or 4 per cell. Some species, in the presence of nutrient deficiency, or as a normal course, produce bi- or quadriflagellate microzoospores which only germinate at low temperatures. The place of these in the life history should be clarified when further species have been studied. Undischarged zoospores can turn into aplanospores that may even germinate *in situ* before liberation. Akinetes) are another means of asexual reproduction.

Sexual reproduction is by means of biflagellate gametes of which 8, 16, 32 or 64 are produced per cell. The plants are dioecious with + or - strains. In *U. flacca,* gametes are produced and fuse under long-day conditions but under short-day conditions they germinate without fusing. Light periodicity is therefore significant and other species of the genus may also be affected.

On germination the zygote either gives rise to 4-16 bodies which function as aplanospores or zoospores, or develops into a 'Codiolum' phase *(U. zonatd).* The species occur most abundantly in winter or spring and optimum conditions would seem to include low temperatures. The genus is well represented in both fresh and salt waters.

Genera closely allied to *Ulothrix* include *Schizomeris* in which some of the cells may divide longitudinally and in which sexual reproduction has not been observed. *Hormidium* species are often included in *Ulothrix* but the filaments have a tendency to fragment, the chloroplast only contains a single pyrenoid and the edges are not lobed, and the zoospores are always biflagellate. *H. flaccidum* is a widely distributed soil alga. In *Binuclearia* the cells are arranged in pairs in a gelatinous filament.

CYLINDROCAPSACEAE: CYLINDROCAPSA

The single genus is found in freshwater and each plant consists of un-branched filaments attached at the base by means of a gelatinous holdfast. When young each thread is composed of a single row of elliptical cells with thick, stratified walls, the whole being enclosed in a tubular sheath.

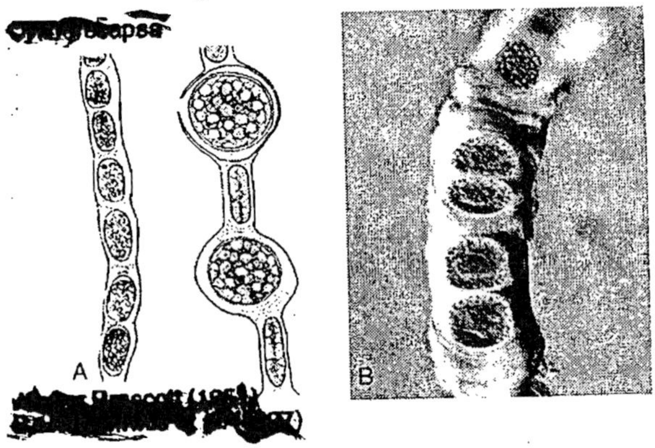

Fig. Cylindrocapsa.

Each cell contains a stellate chloroplast and a single pyrenoid. In older filaments the cells divide longitudinally, usually in pairs, and this suggests how genera such as *Monostroma* may have evolved. Reproduction is more advanced because, apart from bi- or quadriflagellate zoospores, sexual reproduction is oogamous, the plants being monoecious or dioecious. Each antheridium gives rise to two biflagellate antherozoids and each oogonium to a single ovum.

MONOSTROMATACEAE: MONOSTROMA

The thallus originates as a flat, one-layered plate or a small sac, which in most species ruptures very early to give a plate of cells one layer in thickness, the cells often being arranged in groups of two or four. Because there is only one layer the thallus tends to be a paler green and more fragile than the distromatic thalli of *Viva*.

The species of the original genus *Monostroma* are now best treated as falling into three genera: those with the macroscopic leafy thallus alternating with an enlarged zygote or cyst which after some months undergoes meiosis and gives rise to zoospores *(Monostromd)* those with the flattened thallus producing only one kind of swarmer (Af. *oxyspermum)* or with alternation of two kinds of flattened thalli *(Ulvaria)*; at least one species (Af. *zostericold)* with a sporophyte thallus and a small gametophyte disc producing biflagellate gametes.

Bliding places this species in the new genus *Kornmannia,* while Vinogradova removes Af. *oxyspermum* from *Vivaria* and places it in the new genus *Gayralia,* family Gayralia-ceae. A. *latissimum, M. wittrockii* (sensu Moewus), *M. angicava* and Af. *nitidum* are wholly gametophytic and the enlarged zygote produces quadri-flagellate swarmers which give rise to new adults via a flat dwarf thallus *(M. angicavd)* or an erect tube *(M. nitidum).*

Fig. Hitherto Known Life Histories of The Genus Monostroma

In some cases it would appear that not only is there an enlarged zygote but there may also be a dwarf shell-boring 'codiolum' sporophytic phase. In *Ulvaria obscura* vars. *blytii* and *wittrockii* there is apparently a normal alternation of isomorphic gametophytic and sporophytic generations. *M. arcticwn* and *M. tubiforme* only produce biciliate swarmers and so could be either wholly haploid or wholly diploid. Two species, *M. groenlandicum* and *M. lIndaueri* retain the saccate condition throughout their life and probably should be removed to a new genus.

A complete understanding of the original genus will not be possible until the life history of every known species has been thoroughly studied. At the present time no two workers are agreed on the distribution of the various species originally included in the old genus *Monostroma.* It is, however, clear that both life histories and frond development must be considered in any final arrangement.

ULVACEAE: ULVA

The thallus, which characteristically is a flat, expanded, distromatic (two-layered) plate, originates from a single uniseriate filament that subsequently expands by lateral divisions. There is no hollow sac as in *Entero-morpha,* but the two layers may separate towards the margins or at the base as in *U. lima, Enteromorpha stipitata* and *E. flabellata,* so that it becomes difficult to know in what genus to place such transitional species. Typically the flat thallus is unbranched though it can be deeply divided. Branching or proliferating is characteristic of *Enteromorpha* and *Letterstedtia,* but branching has been reported in *U. linearis.*

The young uniseriate filaments are attached by a single cell, but as expansion takes place multinucleate rhizoids grow down from the lower cells and form a basal attachment disc that may persist throughout the winter, new plants arising from it in the spring. Detached fragments are another means of forming new thalli especially in the free-floating forms of quiet waters. The form of the thallus of some species can vary considerably. Plants of *U. lima* found near high water mark in Japan are long and lanceolate and they gradually change to short and broad fronds further down the shore.

Varying exposure times and water loss may be the factors involved here but culture experiments have shown that amino acids and kinetin may also be involved. The plasticity of the species has undoubtedly led in the past to taxonomic confusions, and many of these have yet to be resolved.

The taxon H. *lactuca* has probably included four separate species, *U. lactuca* proper *(sensu strictd)* being anisogamous and with a high latitude distribution, while one of the new taxa, *U. thuretii* is isogamous and a short day plant with a S. European distribution. Similarly a study of *U. linza* from the Mediterranean has shown that the plants differ from those ascribed to the species elsewhere and they have been renamed *U. bertoloni*. It is clear that further work, accompanied by breeding experiments is vitally necessary before the species in this genus are finally established.

Reproduction is by means of zoospores produced in the sporophytic plants and 32 or 64 gametes produced per cell in the gametophytic plants. The latter are generally isogamous, the + and - gametes being liberated from separate plants, but in at least three species anisogamy is known to occur. In some species the gametes may fuse into clumps.

In *U. lobata* two types of 'clump' reaction have been observed. In one type 10-12 gametes fuse together whereas in the other more than 100 gametes are involved. At present there is no explanation for this difference. While the gametes are positively phototactic before fusion, the zygote is negatively phototactic. It has been shown that in certain cases there may be relative sexuality among gametes from different plants, the sex of the older and weaker gametes becoming changed.

Meiosis takes place at zoospore formation and, in most of the species investigated, there is a regular alternation of morphologically similar diploid and haploid generations. In some species only one generation is present, the swarmers invariably developing directly without fusion or without meiosis. C7. *mutabilis* is more complex in that it apparently can exist in the haploid, diploid (gametophytic and sporophytic plants), triploid and tetraploid states. This kind of cytological study is infrequent and there is clearly scope for much further work. In at least one species, *U. lobata,* liberation of gametes is associated with the spring tidal cycles.

The plants occur in saline or fresh water and become particularly abundant when the waters are polluted by organic matter or sewage.

In the allied sub-antarctic genus *Gemina* the plants resemble either *Ulva* or *Enteromorpha.* The cells are normally in pairs and apparently only some cells are capable of reproduction whereas in *Ulva* all cells of the thallus can give rise to swarmers. Similarly, in the genus *Lobata,* the reproductive cells are restricted to the thickened central portion of the thallus.

Papenfuss has argued that *Gemina, Lobata* and *Letterstedtia* should all be merged into the genus *Viva.* Segregation into somatic and germ cells in these genera would seem sufficient justification for retaining them as distinct from *Ulva.* Species of *Letterstedtia* occur in South Africa and New Zealand but Pocock maintains, probably correctly, that the New Zealand *L. petiolata* should be the type of a new genus. *L. insignis* in South Africa is a long strap-shaped thallus with leafy marginal frills and reproductive zones which are separated by purely somatic areas. *L. petiolata* characteristically has a broad rounded blade which arises from several branched, stipe-like portions.

In the genus *Enteromorpha* the plants also start life as uniseriate filaments but they soon become multiseriate and tubular. Like *Viva* many of the species are attached to the substrate by rhizoids, but in *E. nana* there is a basal prostrate disc from which the erect plants arise. This feature, together with small cell size, has been used to create the independent genus *Blidingia,* but because basal discs are known to occur in *E. hendayensis* and *E. tubulosa,* and small cells are a feature of *E. bulbosa,* there seems no justification for retaining the genus *Blidingia* nor for removing it to the family Monostromataceae.

The existence of a basal thallus as well as an erect one forms what is known as *heterotrichy,* a feature characteristic of the Chaetophorales. It seems very likely that the phenomenon of heterotrichy has arisen on a number of occasions in the course of evolution. In addition to the many attached species there are also a few species, occurring especially on salt marshes, which are unattached for the whole or part of their existence.

There is considerable plasticity and hence polymorphism in this genus, as in *Viva,* so that taxonomists have either accepted a very large number of species or else has described few species with many forms. Further work, including culture and crossing experiments will be necessary before a final solution is reached. Taxonomically one can conveniently divide the species into groups which appear to have common features: the *compressa-intestinalis* group, the *clathrata*, the *nana*, the *procera* and the *prolifera* groups. Sliding also recognises *torta*, *jugoslavica*, *flexuosa* and *linza* groups, but he removes the *nana* group to *Blidingia*.

Growth in length of the thallus is primarily intercalary, though in the early stages there is also an apical cell that undergoes divisions. In some species the chloroplasts exhibit polarity, occurring normally in the apical portion of the cell. Such polarity is apparently absent in *Ulva* and *Mono-stroma*. When the tubular thallus is damaged by grazers or wave action near the apex the wounded cells put out papillate undergrowths but if the damage occurs near the base, then rhizoids are produced.

In many species there is a dioecious gametophytic generation which produces isogametes and this alternates with a morphologically identical sporophytic generation which produces zoospores. In *E. prolifera*, however, there is alternation between heteromorphic generations, one being simple and the other much branched. It is thought that a similar situation may exist in the case of *E. compressa* var. *australiensis*. Scagel reports that the sporophytic stage of *E. intestinalis* gives rise to tuberculate-like plants which may be identical with the plant known as *Collinsiella tuberculata*. Zygotes produced from gametes of this generation give rise to a disc from which the characteristically tubular *Enteromorpha* plant arises.

This represents an important deviation and should be confirmed. Some species, including *E. nana*, *E. procera*, *E. aragoensis*, *E. adriatica*, are probably wholly diploid as only zoospores are known. In the first-named species meiosis has not been observed and the swarmers have been designated neutrospores. In the other species the swarmers can be either bi- or quadriflagellate and it is clear that cytological study is required to show whether the genus comprises a polyploid series or not.

Anisogamy occurs in *E. intestinalis* where the male gamete is small with only a rudimentary pyrenoid. Parthenogenesis has been recorded for some species, e.g. *E. clathrata,* and this presumably results in new sexual plants. In the reproductive areas some of the cells fail to divide. These are termed rest cells, and later they may germinate and give rise to sporelings which eventually separate from the parent thallus.

Within the genus there are several lines of development including two probable reduction lines, and one such reduction series may be represented by the sequence *E. clathrata* - *E. ralfsii* - *E. salina* - *E. percursa.* The last named species consists usually of only two rows of cells and has been placed in a separate genus *Percursaria.* An isomorphic alternation of generations has been demonstrated, however, and the filaments are said to develop from basal pads so that there seems little justification for generic separation.

Eliding would regard the series as proceeding in the opposite direction with *E. percursa* as the most primitive form, though in a later paper he removes *Percursaria* to a new family, Percursariaceae, which he considers may not even belong in the Ulotrichales. The semi-terrestrial habitat of the species would seem to be against such a view.

CAPSOSIPHONACEAE: CAPSOSIPHON

The family is founded upon the golden-green *Capsosiphon fulvescens* (formerly *Enteromorpha aurea),* but now another species, also golden-green is known in New Zealand. Some authorities do not consider the characters justify a separate family, but on the basis of morphology, reproduction and the unusual brackish water habitat, it can be argued that there is sufficient justification for the treatment adopted.

The cells with gelatinous walls are arranged, often 2-4 enclosed in a common mother cell wall, in horizontal and longitudinal rows but do not separate off into longitudinal areolae as in *Prasiola.* Reproduction commonly takes place by means of akinetes or 4-flagellate zoospores and biflagellate gametes. After fusion a cyst is formed from which 16-32 zoospores emerge. Chihara considers that these facts justify a relationship with the Monostromataceae.

PRASIOLACEAE: PRASIOLA

The young unbranched filament, which is known as the

'hormidium' stage, consists of a single row of cylindrical cells with thick walls which frequently exhibit striations. Later on the cells divide longitudinally and produce a thin expanded thallus, known as the 'schizogonium' stage, which tapers to the base. The cells of the mature expanded thallus are often arranged in fours and possess axile, stellate chloroplasts, while in some species another feature is the presence of short rhizoids that occur in the stalk-like portion or else are produced from the marginal cells.

In the juvenile filament reproduction takes place by means of fragmentation as a result of the death of isolated cells, while in the older, more leafy thallus, 'buds' can arise from the margin. Sometimes the cells produce large, thick-walled akinetes that germinate to form aplanospores from which new plants arise. In *P. japonica,* sexual reproduction is brought about by non-motile macrogametes (16 per cell) and motile microgametes (64 or 128 per cell) that are both produced from the same plant, so that this species at least is monoecious and anisogamous.

Despite the existence of sexual reproduction, multiplication in this species is usually by asexual means. The plants are haploid, but in view of discoveries about *P. stipitata* it is desirable that *P. japonica* be restudied.

In *P. stipitata* the adult plant is at least diploid (12-14 chromosomes), though in view of the count for *P. japonica* it may be tetraploid, and it reproduces by means of aplanospores. Subsequently the cells at the distal periphery (meiospores of Friedmann) undergo reduction division and give rise to cells which are haploid. These cells are monosexual and give rise to either male or female gametes that are anisogamous. The gametophyte segments are therefore attached to the diploid plant.

A similar situation exists in *P. meridionalis* where the segments were formerly regarded as another alga, *Gayella constricta.* In some species swarmers are produced from sporangia and Friedmann considers that these subsequently were retained and finally produced the so-called meiospores and meio-sporophytes of *P. stipitata* and *P. meridionalis,* which must therefore be regarded as highly evolved forms. Ecologically, it has been observed that where *P. stipitata* forms a belt on the shore, wholly diploid plants occupy the upper levels and plants with gametophytic lobes occur lower down.

The shape of the thallus in *P. crispa,* and probably in other species also, varies considerably with the environment. The genus, which is generally absent from the tropics and sub-tropics, is represented by saline, freshwater or terrestrial species, the last named being tolerant towards considerable desiccation. This can be attributed to the lack of vacuoles in the cells and also to a high protoplasmic viscosity. Some authors believe the genus should be separated from the Ulotrichales but this hardly seems warranted, especially in view of Friedmann's interpretation of the development of the P. *stipitata* condition.

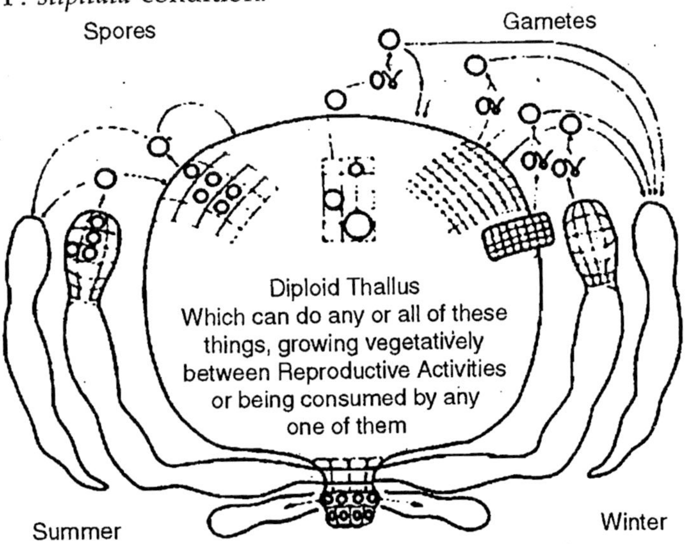

Fig. Life History of Prasiola Meridionalis.

SPHAEROPLEACEAE: SPHAEROPLEA

This genus is widely distributed but is only frequent locally, being most abundant on ground that is periodically flooded by freshwater. The long, free, unbranched filaments consist of elongated coenocytic cells containing up to seventy parietal chloroplasts with pyrenoids.

The chloroplasts occupy the periphery of discs of cytoplasm, the discs being separated from each other by vacuoles, although occasionally they may come together to form a diffuse network. Each disc normally possesses one or two nuclei. In most of the species the septa develop as ingrowths, though in *S. africana* they are replaced by a series of strands which sometimes fail to meet at the centre so that the coenocytes are continuous.

Vegetative reproduction is secured by means of fragmentation, and there is apparently no asexual reproduction though zoospores have been reported in *S. wilmani:* these may, however, be ova developing by parthenogenesis.

In sexual reproduction, although the cells do not change in shape, both oogonia and antheridia are formed singly or in series, the species being either monoecious or dioecious. In the formation of oogonia the annular chloroplasts first become reticulate and then the ova are formed without any nuclear divisions.

In *S. annulina* the ova are non-motile, but in *S. cambrica* the large female gametes are biflagellate and thus represent an advanced anisogamy. In the antheridia the nuclei do undergo division and numerous, elongated, narrow antherozoids are formed which are liberated through small pores, subsequently entering the oogonial cells through similar perforations. The fertilized ovum becomes surrounded by a hyaline membrane inside which two new membranes are laid down and then the first one disappears. The new external membrane is ornamented and the contents become brick red.

Germination stages are only known for some species and the oospores lie dormant for several months or even years. On germination 1-4 biflagellate swarmers (8 occasionally) are produced and these come to rest and then grow into new plants. Sometimes the swarmers do not separate and a four- or eight-flagellate synzoospore may form depending on whether it comprises two or four zoids.

These develop to a four-fold sporeling. In some cases the swarmers from the oospore are completely suppressed and a new filament develops directly, this type of repioduction being known as azoosporic. In the most complete life cycle investigated the motile ova may also develop parthenogenetically after forming a short non-motile structure which divides to produce three new filaments. Various reduction stages from this life cycle occur among the different species. The adult plants are haploid since meiosis occurs at the segmentation of the oospore.

Primitive characters, which seem to be a feature of the genus, are the numerous ova and the entire lack of specialized organs for reproduction. The plant must probably be regarded as an Ulotrichaceous filament which has still retained some primitive features.

In *S. annulina* cells are frequently found with only one or two plastids thus showing a gradation towards *Ulothrix*. There would seem to be very little justification for following those authors who would place the family in the Siphonocladales, even though *S. africana* does have some features characteristic of that order.

4
Euglenophyta

EUGLENACEAE: EUGLENA

This family comprises coloured Euglenoids containing chloroplasts which are usually twisted and elongate. In *E. spirogyra* the pellicle, which underlies a plasmalemma membrane, is composed of a single unit wound in an interlocking helix around the cell. In cross section the unit is composed of grooves and teeth. Arranged parallel with it are a series of mucus bodies, whose function is not exactly known.

Contrary to earlier belief there are two flagella in most of the pigmented forms. The locomotory one (with lateral hairs down one side) is emergent and usually as long as the cell; the other does not emerge from the reservoir. At the base of the locomotory flagellum is a swelling which may be concerned with the photoreception mechanism. The distinctive red eye spot is situated on the edge of the reservoir immediately opposite the flagellar swelling.

Both eye spot and flagellar swelling are absent in *Astasia.* The contractile vacuole (usually only one) is adjacent to the reservoir on the side opposite to the eye spot and discharges into the reservoir. Numerous lenticular chloroplasts are scattered throughout the cell, and pyrenoids may or may not be present. Chromosome counts in algae are not always easy but in E. *spirogyra* the chromosome complement appears to be about 86.

Asexual reproduction is by longitudinal cell division, which may occur either while the cells are motile or resting. Division commences at the anterior end, and where there are non-motile stages a gelatinous sheath may encircle the dividing cell. In some species thick-walled cysts have been recorded, and these germinate to give a single motile cell. Sexual reproduction has not been demonstrated with certainty.

Strructure of A Euglena

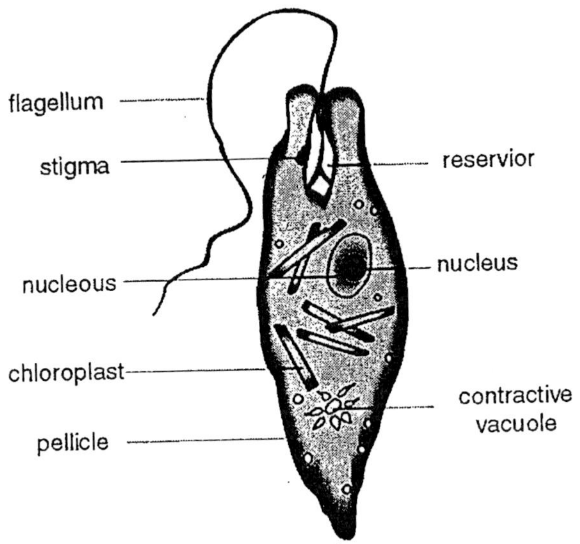

Fig. Euglena

COLACACEAE: COLACIWN

The genus comprises dendroid or palmelloid colonies of immotile cells enclosed in a wall. The species are epizoic and colonies are attached to the host by a gelatinous stalk. Reproduction is by cell division which differs from that of other Euglenoids by occurring at the non-motile stage. Amoeboid stages are known and sometimes the protoplast of a colonial cell may escape and form a free-swimming flagellated individual that will eventually come to rest, lose its flagella and start a new colony..

Phacus is characterised by a flat leaf-like appearance while *Trachelomonas* possesses a distinctive brown cellular sheath. Unicell to the palmelloid state and also through the filamentous to the siphonaceous habit. Morphological development is therefore analogous to that of the Chlorophyceae.

In certain genera in which a cell wall is present, it may be of two overlapping halves. In the filamentous forms the filaments are composed of articulated H pieces, two such pieces comprising the surrounding wall of an individual cell.

The flagellated cells of the Xanthophyceae usually possess two unequal flagella, the short one being of the simple type, the long one being pleurone-matic. It is from this heteromorphic arrangement that the class acquired its early name 'Heterokontae'. The chloroplasts, of which there may be one or more, usually lack pyrenoids, and their yellow-green colour is derived from carotenoids which tend to mask the chlorophyll *a* (the chlorophyll *e*, recorded from *Tribonema bombycinwn* and *Vaucheria haemata* is very probably an *in vitro* product). Asexual reproduction is by cell division, production of motile zoospores, non-motile aplanospores, akinetes, or in some unicellular forms by internal cysts. The statospore or cyst germinates to give two or four zoospores or amoeboid bodies.

All flagellated cells are naked and possess a contractile vacuole system. Sexual reproduction has been recorded for only a few genera, but it is a very distinct feature of the siphonaceous species. The majority of Xanthophyceae are fresh water, sub-aerial (growing on trees, walls etc.) or soil inhabitants and a few are marine.

Recent work has shown that certain species of Xanthophyceae differ in certain important aspects. In the true Xanthophyceae the eyespot is part of the chloroplast. In the following the eyespot is independent of the chloroplast: *Pleurochloris commuta, P. magna* (Pleurochloridaceae), *Polyedriella, Vischeria stellata, V. punctata, Chlorobotrys* (Chlorobotrydaceae) and *Ellipsoidion acuminatum*. Additionally these species differ in that the zoospores possess only one emergent flagellum, and the chloroplast lacks girdle lamellae.

Golgi bodies and the polygonal pyrenoids are not found in motile cells but are present in vegetative cells. Hibberd and Leedale have suggested that these algae should be transferred to a new class, the Eustigmatophyceae, though further work is needed before the validity of this class can be established. It does rise however the important differences within the Xanthophyceae. Other coccoid forms, e.g. *Pleurochloris meiringensis, Botrydiopsis, Bumilleriopsis* are however 'true' Xanthophytes.

CHLORAMOEBALES

The organisms are all naked, motile, flagellated unicells, although temporary conversion to an amoeboid form is not uncommon. The ovoid or pear-shaped cells are uninucleate with one or more discoid chloroplasts. The two flagella are apically inserted, with one or more anterior contractile vacuoles. Reproduction is by cell division of either motile or amoeboid phases or by means of statospores.

HETEROCHLORIS

This genus possesses two parietal chloroplasts and numerous oil droplets. Reversion to the amoeboid form and metabolic movements are quite common. Reproduction is by cell division, although statospores have been reported and in *Chloromeson* they are formed endogenously. The typical Xanthophycean heterokont condition of the flagella is quite noticeable, the difference ranging from 2-3 times the length of the shorter one in *Heterochloris* and *Chloramoeba* to 6-7 times in *Chloromeson*.

RHIZOCHLORIDALES

This order differs from the preceding one in that the vegetative phase is permanently amoeboid, and motile flagellates occur only in the zoospore stage. The majority of the members are uninucleate with one or more chloro-plasts. Reproduction is by cell division. Often the daughter protoplasts do not completely separate but remain connected by their protoplasmic strands.

Chlorarachnion, which contains the biggest species, has uninucleate cells with many chloroplasts and pyrenoids. Nutrition, besides phototrophy, is also holozoic. Cyst formation occurs by rounding off of the amoeboid protoplast and formation of a membrane. The genus *Myxochloris,* which inhabits the cells of *Sphagnum* moss, is multinucleate and has been reported to reproduce by zoospore-liberating cysts as well as by cell division.

HETEROGLOEALES

This order includes those Xanthophyceae possessing a palmelloid habit. The individual cells, each enclosed in a cell wall, are aggregated into palmelloid colonies within a gelatinous matrix, individual cells being capable of reverting to a motile stage.

Reproduction is either by zoospores liberated from the individual cells or by cell division in one or two planes to give two or four daughter cells. Formation of thick-walled resting statocysts may also occur.

In the fresh water *Gloeochloris (gloeo,* jelly; *ehloris,* green) the uninucleate, ellipsoidal cells, containing two or more discoid chloroplasts, are irregularly arranged within the gelatinous envelope. This genus is either free floating or attached to aquatic phanerogams.

MISCHOCOCCALES

This order represents the coccoid tendency in the Xanthophyceae and corresponds to the Chlorococcales in the Chlorophyceae. The morphological similarities are so close that certain species were originally assigned to the Chlorophyceae and *vice versa.* Examination of its pigments has shown that the Xanthophycean *Chlorocloster engadinensis* is typically Chlorophycean and therefore incorrectly classified in the Xanthophyceae. Further biochemical examination of dubiously assigned members is highly desirable and may result in a considerable reshuffling of species.

Unlike the Heterogloëales the vegetative non-motile cells are incapable of reverting to a motile stage. The cells are enclosed in a cell wall which in some species is of the overlapping halves type. Most species are uninucleate and possess one or more chloroplasts within the cell.

Reproduction is either by means of zoospores or aplanospores; strict vegetative cell division appears to be unknown. The majority of species are free floating, or tree, soil or rock dwellers, but epiphytic species are also known. In terms of genera and species, this is numerically, by far the largest order in the class, but, with few exceptions, the species, which are mainly freshwater inhabitants, are of rare occurrence.

PLEUROCH

The family Pleurochloridaceae represents the free-living stage which is an unattached single cell.

Botrydiopsis (like *Botrydium*), is a widely distributed soil alga. The cells tend towards a spherical shape, and the cell wall is apparently in two sections. The cells are uninucleate with numerous chloroplasts (frequently only a few in growing cells), usually without pyrenoids. Reproduction is by means of four, eight or more zoospores of a typical Xanthophyceae form.

Vegetative cells have been known to form aplanospores. These usually germinate to form other vegetative cells, but may pass to a thick-walled resting stage before germinating to zoospores.

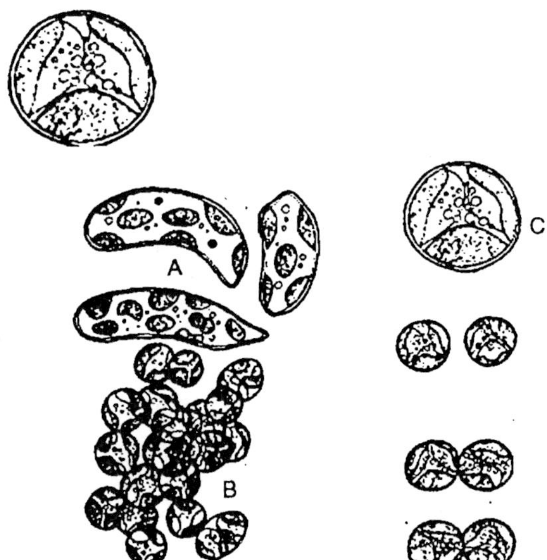

Fig. Xanthophyceae

CHLOROBOTRYDACEAE: CHLOROBOTRYS

The representatives of this family are single cells in free-living colonies enclosed in a gelatinous sheath. *Chlorobotrys* is an acid or alkaline swamp genus. The individual cells are spherical with a few chromatophores. Reproduction is commonly by autospores or thick-walled cysts and zoospore production appears to be the exception rather than the rule.

Botryochloris (Botryochloridaceae) represents a non-mucilaginous, free-living colony of many spherical cells.

CENTRITRACTACEAE

Bumilleriopsis, (like *Bumillerid)* is a typical member with single, free-living, unattached cells that are elongated or ellipsoidal. The cell wall is characteristically and distinctly of two halves. Reproduction in this genus (and family) appears to be predominantly by aplanospores.

MISCHOCOCCACEAE: MISCHOCOCCUS

The members of this family usually form dichotomous dendroid colonies from cells supported on mucilage stalks arranged in a tree-like pattern.

Mischococcus is a freshwater epiphyte. Each mucilage stalk bears at its end a single spherical cell enclosing two chloroplasts. The zoospores are typically Xanthophycean, although those of some species have been reported to possess only one flagellum. When the zoospores settle they give rise to a single spherical cell resting on a mucilage cushion.

This cell then forms a new cell membrane and subsequent secretion of mucilage forms the initial stalk, pushing the cell ahead of it. The original site of the cell and its old cell membrane are marked by a basal swelling. Transverse division of the terminal cell gives two daughter cells that are together supported on a freshly secreted stalk. In some cases only the upper daughter cell secretes the mucilage, in which case the lower cell remains attached to the old stalk, while the other becomes separated by the freshly secreted intervening mucilage.

If longitudinal division occurs the two daughter cells are borne aloft on a common stalk, but at the next cell division the resultant two sets of daughter cells form separate stalks and thus produce a dichotomy. Division into four will give rise to occasional tetrachotomy. Continual repetition of these division processes eventually gives rise to the dendroid colonies. Reproduction also takes place by aplanospores or by vegetative propagation of cells detached from old colonies.

CHARACIOPSIDACEAE: CHARACIOPSIS

The cells, which are epiphytic, solitary or gregarious, vary greatly in shape, even in pure culture; they develop from a short stalk with a basal mucilaginous cushion.

The wall, composed of cellulose and pectins, is in two unequal portions, the smaller upper part forming a lid which is detached at swarmer formation, while in one species the lower part bears internal processes. Although the young cells are uninucleate and contain one or more chloroplasts, the adult cells are multinucleate and contain eight to sixty-four nuclei as well as several chloroplasts. Reproduction is either by means of zoospores (eight to sixty-four per cell) or by means of thick-walled aplano-spores, which in one species are said to give rise to motile gametes, although this is a feature that requires further investigation.

CHLOROTHECIACEAE: OPHIOCYTIUM

The species of the freshwater genus *Ophiocytium* are generally elongated or cylindrical, curved, solitary, multinucleate cells of various shapes, or attached colonies of cells. Dendroid colonies are formed in *O. gracilipes* by liberated zoospores attaching themselves to the rim of the parent cell.

There are many similarities between the Characiopsidaceae and the Chlorotheciaceae, the latter probably representing a specialization of the former. Unlike *Characiopsis* the cells are permanently multinucleate; like that genus, the cell wall *of Ophiocytium* is composed of two very unequal segments.

The lower and larger section of *Ophiocytium* is like an open-ended cylinder with a spine-like projection at the closed end, which serves as the attachment device for the epiphytic species. The open, upper end, through which aplano-spores or zoospores are liberated, is covered by the other section, which acts as a small lid.

In addition to the many nuclei there are numerous chloroplasts, which in certain positions have a distinctive H-shaped appearance. All filamentous Xanthophyceae are grouped in this order in which the majority of species are uniseriate and unbranched. Most possess cell walls with characteristic overlapping halves or H pieces. Asexual reproduction is by means of zoospores, aplanospores or statospores. The latter on germination either form new filaments or liberate a pair of zoospores. Akinetes have also been recorded, and sexual reproduction has been claimed for one species.

The order, which is entirely composed of freshwater forms, is divided into two families, in which the important distinguishing feature is the presence or absence of branching.

TRIBONEMATACEAE: TRIBONEMA

This is a filamentous analogue to a form such as *Microspora* with which it is frequently confused. *T. bombycinum* sometimes appears in sheets covering ponds and pools. The unbranched threads are composed of cells possessing walls of two equal overlapping halves, with the result that the filaments are open-ended and tend to dissociate into H pieces. At cell division a new H piece arises in the centre and the two halves of the cell separate. Each cell contains one nucleus, although *Tribonema bombycinum* may have two, together with two or more parietal chloroplasts. Asexual reproduction is by means of heterokont zoospores which are liberated by separation of the cell wall.

On coming to rest the zoospore elongates and puts out an attachment process, and in this state it resembles *Characiopsis.* Aplanospores and akinetes, which are formed in chains, also act as a means of propagation. Sexual reproduction is apparently very rare; in this some of the motile bodies come to rest first and are surrounded by other motile gametes. Iron bacteria sometimes live within the alga which becomes yellow or brown from ferric or manganese salts.

The branched habit in this order is represented by the family Heteroden-draceae, of which *Heterococcus (hetero,* different; *coccus,* sphere), is an example. This occurs as a discoid filamentous expanse or as single isolated filaments or as non-filamentous cell aggregates. The filaments are each composed of a few short squat cells with one or more parietal chloroplasts. The cell wall appears to be composed of one piece, rather than two H pieces. Reproduction can occur by zoospores, akinetes or by aplanospores. In view of some of its anomalous features, it is possible that this family is not properly included in the Tribonematales.

VAUCHERIALES (HETEROSIPHONALES)

This order includes the Xanthophyceae which are siphonaceous and coenocytic. The plant body is either filamentous or saccate and contains a large number of nuclei and chloroplasts.

Asexual reproduction is by zoospores or aplanospores. Unlike the other Xanthophyceae sexual reproduction (isogamy or oogamy) is quite common in members of this order. There are three or four genera, the majority of species being freshwater or terrestrial, with only a few marine. Two of the the most widely distributed Xanthophyceae genera, *Botrydiwn* and *Vaucheria* are assigned to this order. They are placed in separate families distinguished by their type of thallus and method of reproduction.

BOTRYDIACEAE: BOTRYDIUM

This family is distinguished by a saccate plant body, which reproduces by means of aplanospores or zoospores. The mud-dwelling *Botrydium* is analogous to a form such as *Protosiphon*, the commonest species, *Botrydium granulatum*, being frequently confused with it, especially as these two plants are often associated on areas of drying mud. *B. granulatum* makes its appearance during the warmer part of the year when the green, pear-shaped vesicles are rooted by means of colourless, dichotomously branched rhizoids.

The wall is composed of cellulose and the lining cytoplasm contains numerous nuclei, while the chloroplasts, containing pyrenoids, are confined to the aerial saccate part. The shape of vesicle is influenced by the environment, the shade forms being elongate or club-shaped. In *B. wallrothii* and *B. granulatum* the unbranched vesicle is covered with lime. When the plants are submerged, reproduction takes place by means of numerous zoospores which are set free by gelatinization of the vesicle apex, but when the plants are only wet but not submerged, aplanospores are produced instead.

These are formed by the cytoplasm rounding off and

secreting a wall, and they may be either uni- or multinucleate. They germinate directly to give a new plant. Under dry or drought conditions, the cytoplasm may round off to form a single thick walled cyst (macrocyst), or alternatively several multinucleate thick walled sporocysts or hypnospores. Sometimes the protoplast migrates into the rhizoid and there forms multinucleate hypnospores or rhizocysts. Upon return to favourable conditions these hypnospores germinate to uninucleate zoospores, or if the hypnospore is uninucleate, germination gives rise directly to a new plant.

Sexual reproduction has been claimed for *B. granulatum.* In this species the contents of the cell round off into uninucleate isogametes with typical hetero-kont flagellation. Nuclear fusion takes place after gamete fusion and is followed by the formation of four or eight zoospores. Meiosis is said to occur at division of the zygote nucleus. In *B. granulatum* it is estimated that about 40,000 isogametes are formed in each vesicle, but as the plant is monoecious many fuse either in pairs or threes, rarely fours, before they are liberated. Those that do not fuse, develop parthenogenetically.

VAUCHERIACEAE: VAUCHERIA

In the past this genus has been classified with the siphonaceous Chloro-phyceae in the order Siphonales (Codiales and Caulerpales). Recent work has shown that there are many fundamental differences between *Vaucheria* and the siphonaceous Chlorophyceae and that it is correctly classified in the Xanthophyceae.

The cell walls do not contain native cellulose; chlorophyll *b* is lacking in *Vaucheria* as in all other Xanthophyceae. The discoid chloroplasts, which lack pyrenoids, contain the typical Xanthophycean carotenoids, and lack siphonaxanthin so characteristic of the siphonaceous Chlorophyceae.

The storage reserve is oil. The two flagella present on motile cells are typically Xanthophycean; one is pleuronematic (flimmergeissel or tinsel type) while the other is simple (peitschengeissel or whip-lash type).

Vaucheria is essentially a temperate genus, inhabiting well aerated streams, soil or marine mudflats, where it sometimes forms a greenish covering on the surface. Some species may be slightly lime-encrusted. The plant consists of a colourless, basal rhizoidal portion from which arise green, erect, aerial coenocytic filaments with apical growth and monopodial branching.

All walls are thin, and within the thallus there is a large vacuole that usually runs the length of the cell. The chloroplasts and tiny nuclei are confined to the cell periphery and septa are formed only with reproductive structures or after wounding. Vegetative reproduction is secured through fragmentation, and sexual reproduction is brought about by the well-known, compound, multi-flagellate zoospores (synzoospores) which are produced singly in club-shaped sporangia cut off from the ends of the erect aerial branches.

The chloroplasts and nuclei congregate in the apex of a filament before the septum is laid down and the nuclei then arrange themselves peripherally. Finally, two slightly unequal flagella develop opposite each nucleus and then the zoospore is ready for liberation, a process which is achieved by gelatinization of the sporangium tip.

This compound structure may be regarded as representing a group of biflagellate zoospores which have failed to separate. The syn-zoospore is motile for about fifteen minutes; then it comes to rest and germinates, the first thread often being more or less colourless. Zoospore formation can often be induced by transferring the plants from light to darkness, or from a nutrient solution to distilled water.

Under dry conditions in terrestrial species, thin-walled aplanospores may be formed in sporangia at the ends of short lateral or terminal branches, and if exposed to greater desiccation the threads of the terrestrial forms become septate and rows of cysts are formed to give the 'Gongrosira' stage. When conditions become more favourable these cysts germinate either into new filaments or divide into thin-walled bodies whose contents escape as amoeboid masses, which on becoming sessile germinate into new filaments.

Sexual reproduction is distinctly oogamous, the different species being either monoecious or dioecious. The oogonia are sessile or stalked. Development begins with a bulging in the main thread, and a large mass of colourless cytoplasm followed by many nuclei and chloroplasts moves into the bulge which gradually increases in size. It eventually becomes an oogonium with a single uninucleate egg, which is cut off from the main thread by a transverse system.

Some authors maintain that the extra nuclei which passed into the oogonia are potential gametes which degenerate, while others maintain that the surplus nuclei travel back into the main thread just before formation of the septum. This latter view is considered to be the more likely explanation, although both possibilities may occur dependent upon the species. In the mature oogonium there is a beak, the apex of which gelatinizes, or apical pores through which the antherozoids can pass.

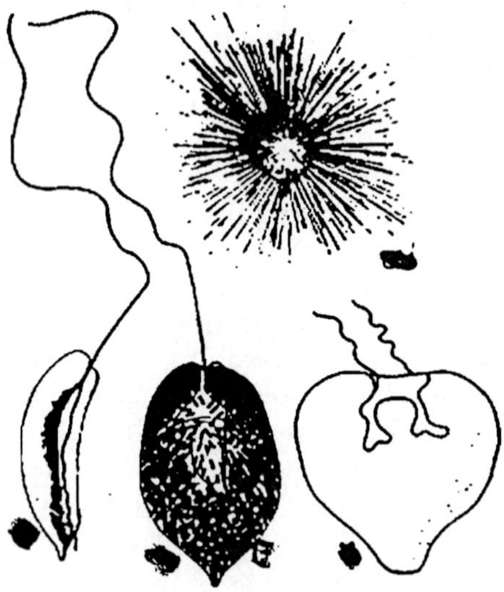

Fig. Gonoyostomum Semen

The antheridia, which are usually stalked, commonly arise close to the oogonia, and in most species their development commences before oogonia formation. The branch bearing the antheridium becomes densely packed with nuclei and a few chloroplasts, so that often the antheridia appear colourless.

When the antheridium, which is often hook-shaped, separates from the main thread by a transverse septum, the contents divide into uninucleate fragments which form pear-shaped antherozooids that possess apically inserted, unequal, heterokont flagella pointing in opposite directions, and are unusual in that the anterior, shorter flagella is of the flimmergeissel type. The antherozooids escape through apical pores and enter the oogonium, although only one antherozoid enters the egg. After fertilization, the zygote secretes a thick wall and remains dormant for some months before it germinates to give rise to a new filament.

The latest evidence indicates that meiosis occurs at germination of the zygote, thus indicating that the adult plant is haploid.

Under certain environmental conditions (e.g. viscous agar medium or increasing salt concentrations due to evaporation) the cells may become spherical, cyst-like or amoeboid with metabolic movements. Under other conditions the cells may become aggregated in a gelatinous matrix and then they sink to the bottom of the pool or pond.

In both *Gonyostomum semen* and *Vacuolaria virescens* encystment can occur, whereby the cells become spherical and surrounded by a sheath: this represents a different phenomenon. Cell division is by longitudinal fission while in a somewhat amoeboid form.

Sexual reproduction has been recorded for *Vacuolaria virescens* and *Hornellia marina* (possibly *Chattonella subsalsd)*. In the latter case two flagellated individuals fuse and lose their flagella. *Hornellia marina* is the sole monotypic marine genus and occurs off the west coast of India, often in sufficient quantity to cause a green discolouration in the sea.

5
Phaeophyta

The Laminariales are principally temperate, the bulk of the species being confined to colder waters, and there are, {^particular, a number of monotypic genera confined to the North Pacific.CThe thallus, representing the large, conspicuous sporophytic generation, is nearly always bilaterally symmetrical with an intercalary growing zone, while the gametophytes are microscopic ectocarpoid filaments the sporophytes reproduce by means of zoospores formed in unilocular sporangia, commonly arranged in sori with paraphyses, while the gametophytes reproduce by memis of ova and antherozoids that are borne on separate microscopic plants. The gametophytes only mature under low water temperatures. The number of zoospores produced per sporangium ranges from 16 in *Chorda* to 128 in *Saccorhiza.*

Cohere is considerable variation in habit between the sporophytes of the different genera. They rangeJrom the long whip-like thallus of *Chorda* to the giant *Macrocystm.* Between these extremes are found the sea palm, *Postelsia* and me sea fern, *Thalassiophyllum,*(Jn the northern hemisphere *Laminaria* is a typical representative of the gem and in the southern, *Ecklonia,* while *Macrocystis* is a useful teaching representative in the Pacific and in South Africa.

The genera *Saccorhiza* and *Alaria* are of interest because the fronds possess shallow conceptacles or cryptostomata and resemble those found in the Fucales, those in *Saccorhiza* even possessing the typical tuft of hairs. There are five families in the order:

a. Chordaceae: the plants are long, whip-like and with deciduous hairs. It is a monotypic family.
b. Laminariaceae: the species either possess a simple blade or one that is segmented longitudinally, but

the splits do not extend into the transition zone so that there is only one intercalary meristem.

c. Himantothallaceae: branching is distichous and lateral from the primary stipe. There is a unique internal assimilatory tissue associated with conducting vessels. Reproduction is more like that of *Desmarestia* than *Laminaria*.

d. Lessoniaceae: the splits of the primary blade extend into the transitionzone so that each segment has a stalk and an intercalary meristem.

e. Alariaceae: plants have special lateral sporophylls.

CHORDACEAE: CHORDA

The long whip-like thallus, which is clothed in summer with mucilage hairs, arises from a small basal disc on the growing region situated just above the holdfast. The hollow fronds are simple with diaphragms at intervals, the thallus being like that of a multiseptate cable whose construction is derived from the multiaxial chordarian type by further segmentation of descending hyphae to form a pseudo-parenchyma.

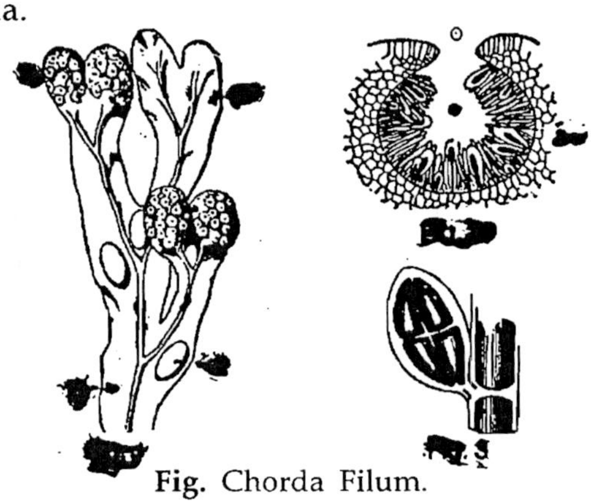

Fig. Chorda Filum.

The epidermal layer is ultimately clothed with sporangia, paraphyses and deciduous mucilage hairs, while the central cells become elongated and support the filaments that go to form the diaphragm. The zoospores on germination give rise to small filamentous gametophytes, the male plants being composed of small cells, each with two to four chloroplasts, and the female of larger cells with more numerous chloroplasts.

The gametangia are borne laterally or terminally on short branches, but the plants do not become fertile for at least three months or more often six months after their formation. After fertilization the zygote remains attached to the wall of the oogonium. The macroscopic plant is an annual.

LAMINARIACEAE: LAWINARIA

This genus has a very wide distribution in the waters of the north temperate and Arctic zones. The expanded lamina has no mid-rib and is borne on a stipe that arises from a basal holdfast which can vary greatly in form. In *L. angustata* populations may have either ovate or lanceolate laminae though sporelings of the former type frequently disappear at an early stage of their existence.

The simplest transition area from stipe to lamina is quite plain, but folds, ribs or callosities are also found in that position, which is the region of intercalary growth. In *Laminaria sinclairii* three types of growth can be recognized, all of them confined to the stipe, and it is also possible to find all three processes taking place in one individual:

1. The ordinary growth and extension of the blade during the growing season. This hardly merits the description of continuous physiological re-generation given to it by Setchell unless the concept of regeneration is to have a wider significance.
2. Periodic physiological regeneration which represents the annual formation of a new blade. The transition area bulges, due to new growth in the medulla and inner cortex, and then ruptures from the pressure, to leave the frayed ends of the non-growing outer cortex with collars, the upper one of which rapidly wears away. After the rupture the new cells of the medulla and inner cortex elongate rapidly. The failure of the outer cortex to grow is probably associated with the proximity of the inner cortical cells to the medullary hyphae where they can monopolize all the growing materials, thus cutting off any supply to the outer cortex. There may, of course, be other factors involved.

3. Restorative regeneration whereby branches arise from wounded surfaces, the same tissues being involved as in process.

A detailed study of growth in *L. saccharina* by Parke has shown that in this species the longevity and rate of growth of the sporophyte depend on the season of zygote germination, depth in the sea and type of habitat. Thus winter sporophytes rarely attain maturity and the bulk of the populations in Great Britain originate in the spring, except on sheltered coasts where summer plants thrive equally well.

The total life span of this species and also *of L.japonica* rarely exceeds three years but that of some others may be longer. Growth is seasonal, more rapid growth occurring between January and June and slower growth afterwards. In the Barents Sea the blade of *L. digitata* only grows from spring to July whereas the stipe grows all summer. In Norway growth of the new lamina commences in December as the waters are rather warmer than those of the Barents Sea.

Fig. Laminaria.

The attachment 'roots' or haptera, however, attain their maximum growth in the autumn. In *L. saccharina, L. hyperborea* and L. *digitata* the seasonal changes in growth rate are indicated in the stipe by alternate zones or 'rings' of lighter and darker tissue, the darker being formed during the periods of slow growth. In L. *saccharina* the maximum height and greatest growth is achieved during the second rapid growing season. In L. *digitata* four annual growth rings appears to be the maximum but most plants usually have less. In *L. hyperborea* the number of growth rings is usually seven but Scottish plants apparently live longer and for that reason form taller underwater 'forests' than the same species farther south.

Making use of transplants and measuring growth by means of holes punched in the lamina, Sundene has shown that the shape of the frond in L.*digitata* is greatly influenced by growth conditions, and in particular by temperature. At the period of maximum growth the mean growth increment is equivalent to about 14 cm per 30 days. Growth takes place all over the thallus but the greatest growth occurs in the basal meristematic region. The width of the thallus seems to depend on the degree of protection so *that forma stenophylla* is simply an expression of extreme wave action. The splitting of the fronds in this species is correlated with strong water movement.

Many of the species are used as food by the Russians, Chinese and Japanese. Furthermore, these kelps, as they are called, were formerly valuable as a source of iodine and as a potassium fertilizer. Their present main commercial use is as a source of the important alginates.

Morphologically both lamina and stipe are divided into four regions. On the outside there is the actively dividing meristoderm or limiting layer, and inside there is an outer and an inner cortex and then a mass of interwoven threads that form the central medulla. The young sporeling is at first a uniseriate filament, but as a result of cell divisions it becomes a flat mono-stromatic plate. Rhizoids emerge from the basal cell and the thallus then becomes distromatic, the periclinal divisions that are responsible starting at the basal end.

When the holdfast, stipe and blade have been differentiated,, further growth is restricted to the intercalary meristem at the base of the blade. When a distromatic lamina has been produced a central group of cells, which form the primary tubes of the medulla, are cut off and separate the two outside layers. Although the origin of the medulla in the stipe is clear, it is not so for the lamina. The intercalary growing zone forms a ring around the base of the lamina, the central tissues not being meristematic.

Later the primary limiting layer in the stipe is replaced by a secondary meristem which is four to eight cells deep in the cortex. This is responsible for the subsequent growth in thickness by production of secondary annual cortical tissue.

The medulla consists of primary medullary tubes, and two types of lateral connecting branches which are the connecting threads and the hyphae. The former arise first in the course of development as papilloid outgrowths from individual cells. They meet and fuse at their tips and so give rise to a system of cross connections. With further growth the cells divide and elongate, but even when mature they are composed of relatively few cells.

The hyphae, which arise later as branches of small cells cut off from the original vertical cells, remain free but may branch. Ultimately they contain numerous cells which subsequently elongate very considerably. One of the most characteristic features of the genus, and indeed of the order, is the presence in the medulla of 'trumpet-hyphae', which arise as modifications of some of the cells, usually in the connecting threads, more rarely in the hyphae.

At a transverse cell wall the ends of both cells swell out to form bulbs, the upper bulb always being larger. The transverse wall is perforated to form a sieve plate, and a callus develops on each side, both callus and sieve plate being traversed by protoplasmic strands. Apart from the sieve plates the trumpet hyphae also possess spiral thickenings which appear as striations. The function of these trumpet hyphae is still unknown. In some species many of the other cells also contain pits with a thin membrane across the opening and these presumably facilitate the transport of storage reserves.

Most of the genera possess systems of anastomosing mucilage ducts lined with secretory cells. In *L. hyperborea* these are confined to the stipe and in *L. digitata* to the frond, but in *L. saccharina* they occur in both. They arise lysigenously through an internal splitting of the thallus caused by cell disintegration; this is followed by a differential growth so that the canals become more and more submerged in the thallus. The haptera have an apical growth and differ from the rest of the thallus in that there are no connecting hyphae nor is there any medulla.

The sporangia and paraphyses are borne in irregular or more or less regular sori on both sides of the lamina. They arise from superficial cells that grow out from the surface.

These divide tangentially into a basal cgll and paraphysis initial, and later a sporangium is cut off from the basal cell. In *L. saccharina* the sporophytes only reach maturity at the end of their second year, and for sori to develop the distal tissue of the frond must be at least six months old.

The actual number of reproducing plants in a population and the duration of the reproductive period depend largely on the numbers of the different season and age groups making up the population. In general *L. saccharina* in Great Britain fruits in summer, L. *hyperborea* in winter, and *L. digitata* from April to November with maxima in spring and autumn.

The zoospores liberated from the unilocular sporangia germinate to form minute gametophytes. On germination they first put out a tube that terminates in a bulbous enlargement into which the contents of the zoospore migrate. There the nucleus divides and one daughter nucleus passes into the tube while the other is said to degenerate, but at present the significance of this phenomenon is obscure the gametophytes show a great deal of variation in shape and size, the male gametophyte being the smaller throughout as it is built of smaller cells.

The male gametophyte always consists of more than three cells whereas the female may consist of only one cell, the oogonium; both however may become branched filaments. The antheridia arise terminally or intercalarily and produce only one antherozoid, whereas any cell of the female gametophyte may function as an oogonium. The male gametophyte degenerates after the gametes are shed but the fertilized oogonium is normally retained on the female gametophyte which persists during the early growth stage of the sporophyte.

(Reproductive organs are only formed at low temperatures such as, 2°C-6°C, and above 12°C-16°C they are rarely produced; this fact accounts for their temperate and arctic distributional is also known that the eggs may develop parthenogenetically to give a haploid sporophyte of irregular shape.

It has been possible to produce hybrids between *L. japonica* and *L. religiosa* or *L. ochotensis* and also between *L. religiosa* and *L. ochotensis*. Crosses between other species produced abnormal sporophytes of types not found in nature. It has also been possible to produce hybrids between *Laminaria* and *A laria.*

ffhe young sporophyte first produces numerous rhizoids of limited growth, but these are later covered by a disc-shaped expansion, in older plants this disc is replaced by the haptera which arise as papillate outgrowths from-the stipe.

Saccorhizapolyschides differs from *Laminaria* in several important respects:

The persistent lamina arises from a flat compressed stipe with wavy edges which, as a result of unequal growth, are twisted through an angle of 180° near the base.

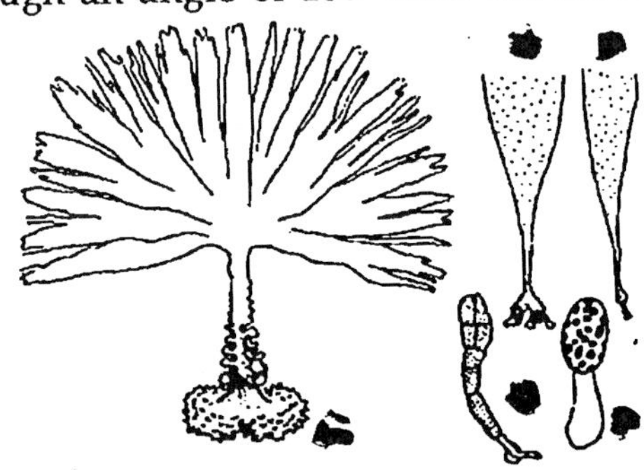

Fig. Saccorhiza Polyschides

It differs from *Laminaria* in lacking cortical mucilage ducts and sieve-tube like elements. The young sporophyte is attached at first by a small cushion-like disc, but later a warty expansion, the rhizogen, develops above it and forms a bulbous outgrowth which bends over and attaches itself to the substrate by means of haptera. As a result of the development of this adult holdfast the juvenile disc may be lifted completely off the substratum.

The development of the sporophyte to maturity in both known species requires only one year so that the plants are mostly true annuals. Some plants occasionally overwinter, and the bulb certainly does. *Saccorhiza polyschides* is found on the Atlantic coasts of North and West Europe whereas *S. dermatodea* is circumpolar.

In the Pacific genus, the perennial sporophyte is apparently composed of a spirally twisted, fan-shaped lamina unrolling from a one-sided scroll without any mid-rib. A study of the embryonal stages, however, shows that the young plant is flat and bilaterally symmetrical. The edges then curl up and the plant tears down the centre giving rise to two lateral scrolls each unrolling from a thickened outer margin; one of the scrolls soon ceases to develop and the mature plant only possesses one scroll borne on a solid bifid stipe with the vestigial scroll on one of the branches. Slitting is represented by rows of small holes which start to develop when the first tear has taken place.

HIMANTOTHALLACEAE: PHYLLOGIGAS

This is a monotypic family with but a single genus and species, *P. grandi-folius,* though other workers have segregated the different growth forms into separate species. It is a sublittoral alga of the Antarctic continent. In young plants the branching is from the primary stipe and is distichous. In older plants only stumps of branches may be left. As the plants age an abundance of hapteres are produced from the basal part of both stipe branches so that a complex rhizome-like body is formed.

Fig. Phyllogigas Grandifolius

When only two branches of almost equal size are left, branching then appears to be dicho-tomous. In many plants the stipe is spirally twisted. Growth is by means of an intercalary growth zone and there is also a superficial meristem. The genus differs from other members of the Laminariales in that there are large, elongated, internal conducting vessels each surrounded by a sheath of assimilatory cells filled with plastids. The reproductive bodies are also distinct. They comprise unilocular sporangia with a corona-like structure at the apex, mixed with sterile, two-celled paraphyses, the whole being part of a modified epidermis.

In this respect the plant shows a resemblance to *Desmarestia*. Ultimately, as Neushul suggests, it may be necessary to place this genus in a separate order.

LESSONIACEAE: LESSONIA

This family contains some of the largest known algae. Species of *Lessonia* form 'forests' in relatively deep waters off the shores bounding the southern Pacific. The stipe *ofL.flavicans* is extremely stout and rigid, 2-3 m. long and sometimes 15 cm or more in diameter, the additional growth being due to a deeper seated secondary cortical meristem.

It appears to be more-or-less regularly branched in a dichotomous fashion, a feature which is brought about by the lamina being slit down successively to the intercalary growing region, each successive segment developing into a new lamina with its own portion of stipe.

Two interesting genera are *Nereocystis* and *Pelagophycus.* Both arise from small holdfasts but attain stipe lengths from 15 m. The upper 2 to 3 m of the stipe is expanded and hollow, forming a long thick-walled (1-2 cm) pneumatocyst that terminates in a bulbous sphere.

Fig. Lessonia

In *Nereocystis* the maximum diameter is about 15 cm but in large *Pelagophycus porra* diameters of 30 cm may be reached. The total volume may approach 4 dm^3 and the gas mixture inside is less than atmospheric. The gas contains oxygen, nitrogen, carbon dioxide and up to 12% carbon monoxide (usually 1-5%).

The top of the pneumatocyst subtends one *(Pelagophycus)* or four *(Nereocystis)* extensions of the stipe. These dichotomise in a pattern characteristic for each genus, one branch of each dichotomy bearing one blade. The blades in *Pelagophycus* are very large, up to 10m long and 1-5-3 m wide. The blades of *Nereocystis* are more numerous (up to 64 in number) and smaller. The characteristic branching pattern accounts for the names Elk-horn Kelp *(Pelagophycus)* and ribbon Kelp or sea-otters cabbage *(Nereocystis)*.

The three known species of *Pelagophycus* represent a morphological series and Parker & Bleck have argued that *P. gigantea* is an autopolyploid *ofP.porra,* while *P. intermedia* is a hybrid. A third fossil genus *Julescraneia* has recently been described which is morphologically intermediate between *Nereocystis* and *Pelagophycus.*

LESSONIACEAE: MACROCYSTIS

The perennial fronds may be 60 m long. The upper part of the plant floats on the water surface, kept there by the basal bladders of the leaves. In the juvenile plant the stipe is simple and solid, but later it branches one to three times in a dichotomous fashion, although ultimately the branching becomes unilateral and sympodial. The growing region on each branch is basally situated in the terminal blade, and along the edge it is continuous with the stipe, and it is in the former area that splitting takes place to form the individual laminae.

The splitting is brought about by local gelatinization of the inner and middle cortex together with a cessation of growth in the epidermal area; this forces the adjacent tissues into the gelatinized areas until finally the epidermis is ruptured. One of the two segments formed in this splitting process stops growing while the other, the outer, continues growth.

The internal structure of the stipe and fronds, with the

exception of the 'sieve' tubes, is typical of the order as described under *Laminaria*. The rate of growth of this plant, during the growing season, is quite spectacular. On the Pacific coast of North America the terminal portion grows at an average rate of 7-1 cm per day, which works out at about 1-1J blades daily. At a depth of 20 m whole fronds can grow as much as 45 cm per day, the most rapid plant growth known. The amount of organic matter accumulated in the growing tip is equivalent to that produced and translocated from 35 blades immediately below.

There are very many more blades than this in the plant during most of its life, so that the remaining surplus of photosynthesized material must be translocated to the stipe and haptera and stored as reserve. Tagging experiments have shown that the average age of an individual frond with its blades is about six months, so that the rate of turnover is considerable.

Macrocystis and *Nereocystis* possess 'sieve' tubes in the cortex as well as trumpet hyphae in the medulla. The 'sieve' tube callus has been found to resemble callose from the sieve plates of vascular land plants. The plates are penetrated by strands of material and there must be some kind of mass flow as Parker has shown that the rate of movement of photosynthate is 65-78 cm per hour. Neighbouring tubes are connected to one another by lateral filaments, the cross walls of which are also perforated.

Reproduction is by means of zoospores produced on sporophylls which are generally dichotomously branched and with or without a bulb, near the base of the plant. Although two kinds of zoospore have been recorded, recent work has failed to substantiate this finding. Enormous numbers of zoospores are produced, the actual number being related to the size of the frond. Each sporangium produces 32 zoospores and on a large frond as many as 700,000 zoospores may be liberated per square millimetre of frond (both surfaces). On germination the zoospores give rise to the characteristic laminarian game-tophytes, which may live for six months or more.

Their form is determined by nutrient supply, bacterial growth, temperature and so on and this leads to presence or absence of sex organs, extreme branching or very compact branching. The ova are normally fertilized in the sea, and though occasionally a single ovum may be fertilized while still in the oogonium, this can only be regarded as an abnormality. Successful growth and establishment can be greatly affected by grazing organisms, especially sea urchins, and abalones *(Haliotis)*.

ALARIACEAE: ECKLONIA

This is primarily a southern-hemisphere genus. There is a short, solid unbranched stipe which is attached to the rocks by means of the usual haptera. In the young plant the stipe expands into a flattened frond which in most species is initially broadly dentate.

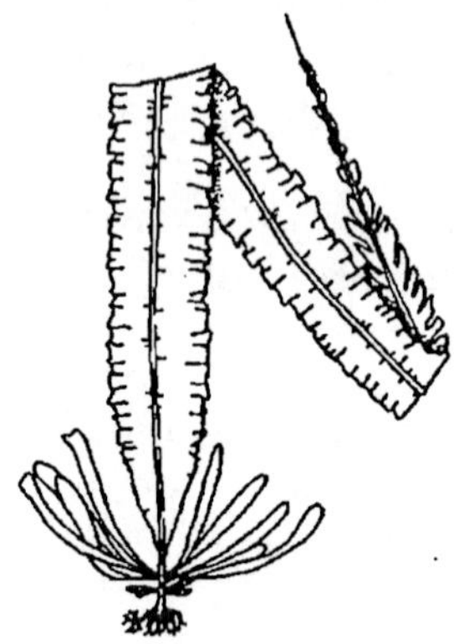

Fig. Alaria Esculenta.

Later the terminal portion disappears and the basal portion produces numerous lateral outgrowths or sporophylls. In its internal morphology the structure is comparable to that described for *Laminaria.* There is evidence that the shape of the frond in *E. radiata* is determined by the degree of water movement so that var. *richardiana,* which only partially leaves the juvenile condition, is an ecological form associated with strong water movement. The genus tolerates rather higher sea temperatures than do other genera of the Laminariales and is therefore found at the north of New Zealand; *E. kurome* also penetrates south into the warmer waters of Japan. Hybrids have been successfully produced between this species and species of*Laminaria.*

In the northern hemisphere the Alariaceae is represented by the widely distributed genus *Alaria*. There is a short, solid, unbranched stipe, naked below, with an intercalary growing zone that allows for continual renewal, while above the growing region the stipe expands into a flattened rachis which each year bears a fresh crop of marginal rows of sporophylls.

The frond terminates in an expanded, sterile lamina which has a well-marked mid-rib and is produced annually. In addition to intercalary growth there is also marginal growth that imparts a wavy appearance to the terminal frond. The sterile frond bears so-called cryptostomata, although these are barely more than tufts of hairs arising in slight depressions. Growth of *A. esculenta* is not good above a temperature of 16° C and the sporophyte dies above 20° C.

The distribution of this species is therefore restricted to waters with a mean August sea-temperature of less than 16° C. Below this temperature, growth of the blade is also reduced if the salinity is lowered. The plant does not therefore normally occur around estuaries. The gametophytes are protonemal in form, simple or sparingly branched.

The ovum is fertilized on emergence from the pore of the oogonium and the young sporophyte develops *in situ* without the characteristic early appearance of a holdfast. An extension of the *Alaria-type* can be seen in the North Pacific American genus, *Egregia*. Each branch is strap-shaped and bears three types of outgrowths: ligulate sterile leaflets, small fertile leaflets, and conspicuous stipitate bladders. The female gametophyte is composed of one or two large cells, while the male is composed of smaller ones.

ASCOSEIRACEAE: ASCOSEIRA

In the single species, *A. mirabilis*, young plants have a single blade and there is a short stipe arising from a pad-like hold-fast. The single blade splits down the middle at an early stage and this primary split is followed by more or less symmetrical splitting of both halves. Splits originate from the tips of the blades. Old plants may reach nearly 2 m. in length. Blades of large plants show constrictions and it is thought that these may indicate seasonal regeneration.

Structually the epidermis or limiting layer and the cortical cells beneath are very comparable to those of *Fucus*. The medulla possesses elongated tubes, which probably have a conducting function and these are surrounded by thin-walled elongated cells and hyphae which can be united by connecting branches, very much as in the medulla of *Laminaria*. The conducting cells are surrounded by a ring of small assimilatory cells containing brown plastids. In this respect *Ascoseira* is like *Phyllogigas*.

The laminae have the same structure as the stipe, but in older, well-developed plants there are, in addition, conceptacles characteristic of the Fucales and comparable to those found in *Notheia* and *Splachnidium*. Each conceptacle has a narrow canal which opens to the surface and in the early stages is filled with hairs that have developed from the floor.

Eventually chains of cells arise from special cells, 'sporangial cells', in the boundary walls, and when ripe each cell of the chain divides to give eight smaller cells, so that chains of eight-celled bodies are formed. The 'sporangial cells' that give rise to the fertile chains could be interpreted as unilocular sporangia that germinate *in situ* to produce the fertile chains.

The exact nature of the fertile bodies remains unknown. In very old plants white patches can be seen on the thallus and these contain what appear to be unilocular sporangia that give rise to discrete bodies. This may be some form of accessory reproduction.

DURVILLEACEAE: DURVILLEA

The diploid plant is dark olive-brown or black in colour and looks very much like a *Laminaria*. This applies to all four species (five if *Sarcophycus potatorum* of Tasmania is included). The large, solid stipe arises from a scutate holdfast and very soon passes into a flat, expanded, fan-shaped lamina, which later splits into segments although no definite appendages are produced from the frond.

The New Zealand *D. willana* is characterized, however, by lateral fronds proliferating from the stout stipe. The attachment disc bears a resemblance to the primary disc found in some members of the Laminariales, and it is formed in much the same way by tangential divisions of the outer layer or meristoderm. The growing region is at the margin.

Internally the thallus possesses small hyphae which anastomose and fuse with one another just as in the medulla of *Laminaria*. These hyphae extend right up to the apex, thus differing from genera such as *Fucus*. There is also a mechanism for splitting the thallus as there are thallus-splitting mechanisms in the Laminariales. Before splitting, new cortical tissue is formed.

In the most widespread species, *D. antarctica*, the adult lamina is characterized by large, internal air spaces separated from one another by septa. These air spaces are thought to result from an internal splitting process which may or may not be associated with degeneration of the tissues. In this genus, growth is localized at the tips of the laminae, the surface layer forming the primary meristematic layer.

The ends of the older laminae become frayed and broken off by wave action, while the holdfast may attain a diameter of 60 cm through the annual addition of new tissue. If this secondary growth did not occur the plant would soon be torn from its moorings because the holdfast is continuously riddled with holes made by boring molluscs. The oogonia and antheridia, which are borne in conceptacles on different plants, occur over the whole of the lamina.

The oogonia (macrosporangia) are usually borne on branched hairs, whereas in all other genera they are borne directly on the wall. This condition in *Durvillea* is comparable to the antheridial structure generally in the Fucales and must be regarded as the primitive condition. The ripe oogonium contains four ova. The plant is known as the 'bull kelp' and forms submarine forests in deep waters off New Zealand, the Antarctic islands and the southern part of South America.

FUCALES

The diploid plants are wholly dominant in the life cycle but although dip-loid there is no apparent asexual reproduction as the plants always reproduce by means of ova and antherozoids. In most genera it is likely that the plants exist over more than one season: this has been clearly established for species of *Fucus, Ascophyllum,* and *Marginariella urvilliana.*

There is considerable tissue differentiation, and in their external features the plants exhibit much more variation than is to be found in the Laminariales. The thallus is normally differentiated into a discoid holdfast, stipe and frond, and growth in length is due to the activity of one or more apical cells. The genus *Bifurcaria* is anomalous in that the discoid holdfast is replaced by a basal rhizome.

Morphologically, an external meristoderm, cortex and inner medulla can be recognized. In the genera commonly occurring in the intertidal regions the walls of the central or medullary cells become heavily gelatinized and there is a greater profusion of longitudinally running hyphae; *Hormosira* is an apparent exception to this in that the swollen segments become hollow. In the genus *Halidrys* sieve plates occur on the walls of the medullary cells and there is an indication of similar plates in the lateral medullary connections of *Bifurcaria*. Scattered over the surface are minute cavities, the *cryptostomata,* which bear sterile hairs or paraphyses.

Reproduction is by means of oogonia and antheridia borne in conceptacles, which are derived from enlarged cryptostomata and are localized in regions known as receptacles. Some workers consider that the structures called oogonia and antheridia are really macro- and microsporangia producing mega- and microspores which germinate before they are liberated from the sporangium, so that while the reproductive bodies originate as spores, nevertheless the liberated products are gametes. This view is significant on a phylogenetic basis. In the primitive condition eight ova are produced in each oogonium and 64 antherozoids in each antheridium.

Meiosis takes place during the first two divisions in the formation of microspores, and as there is often a pause after the second division the first four nuclei have been regarded as the functional microspores, each of which subsequently undergoes four mitotic divisions so that they can be said to germinate to a 16-celled gametophyte where each cell functions as an antherozoid.

In the macro-sporangium the first four nuclei formed have been regarded as the functional megaspores, and each of these is considered to germinate subsequently to a two-celled female gametophyte where each cell functions as an ovum. In those species where less than eight mature ova are produced it must be assumed that some of the megaspores abort or else do not develop.

If this is the correct interpretation, and it would seem to be more satisfactory than any other theory in comparison with other members of the Phaeophy-ceae, then not only is there a cytological alternation of generations but there is also a morphological alternation, although the sexual generation is even further reduced from that found in the Laminariales. This would really form a basis for placing the Fucales in the Heterogeneratae.

The alternative interpretation is that the sexual generation has been completely suppressed and is solely represented by the gametes, so that while there is a cytological alternation of generations there is only one morphological generation.

The flask-shaped depressions of the thallus (conceptacles), are lined with sterile hairs or paraphyses, and open to the surface by means of an ostiole. The plants of the different species may be dioecious or monoecious. In a number of the fucoids the ova, when they are extruded through the ostiole, remain attached by means of a thin gelatinous stalk.

These stalk-forming fucoids, e.g. *Scytothalia, Marginariella, Sargassum, Bifurcaria,* have been studied in some detail and they have certain features in common:

1. The stalk is formed from the mesochiton or middle wall of the oogonium;
2. They are inhabitants of deep water or pools near low water of spring tides *(Bifurcaria brassicaeformis* of South Africa is an exception);
3. The conceptacles are unisexual;
4. They have only one ovum per oogonium *(Bifurcariopsis capensis* is an exception);
5. Maturation takes place at a late stage;
6. The fertilized ovum begins development within the jelly of the attachment stalk.
7. Conceptacles are borne on specialized laterals;
8. The gelatinous stalks are hollow;

9. They are endemic genera or species, of restricted distribution.

The number of primary rhizoids in the embryo depends first on the species and also on the size of the rhizoidal cell, which in turn bears a relation firstly to the size of the egg, and secondly to the complexity of the thallus. On this basis a series of increasing embryonal complexity may be traced, e.g. *Fucus — Ascophyllum — Pelvetia — Cystoseira — Sargassum.* The eight-egg condition has generally been regarded as the more primitive.

On the other hand a study of the antherozoids, using the electron microscope suggests that *Cystoseira* is more primitive than Fucus. Views on evolution within the order will obviously need re-examining. The Fucales are classified into four groups, the classification being based primarily upon the structure of the apical growing cell or cells:

1. Fuco-Ascophyllae Growth is determined in the adult stage by one four-sided apical cell. Recently it has been found that in *Marginariella urvilliana* the apical growing cell is three-sided in the adult condition. The apical growing cell of the salt marsh fucoids has also been shown to persist in the three-sided condition. Despite these exceptions, however, the basis of the classification remains very useful.
2. Loriformes (Himanthaliaceae): Growth is due to *one three-sided apical* cell which gives rise to a long and repeatedly forked, whip-like thallus.
3. Cystoseiro-Sargassaceae: The apical cell is again *three-sided but* there is copious branching which results in bilateral, radial and bilaterally-radial thalli.
4. Anomalae: Composed of one genus, *Hormosira,* which is confined to the Antipodes. Growth is brought about by a group of cells, not a single cell.

In most of the Fucales the apical cell is sunk in an apical pit or groove. In the Fucaceae proper, this groove is usually parallel to the plane of flattening of the thallus, but in *Halidrys* and *Himanthalia* it is at right angles to it. In the adult the hairs in the young apical pit are deciduous, but in *Himanthalia, Bifurcaria* and *Xiphophora* hairs can be found growing out from the apical groove of mature plants.

FUCACEAE: FUCUS

This genus contains a number of species that are widely scattered in the northern hemisphere, many of them exhibiting a wide range of form with numerous so-called varieties. When two or more species occur in the same area they are generally present in different zones on the shore, probably dependent on the degree of desiccation that they can tolerate. The plants are attached by means of a basal disc and there is usually a short stalk, which continues to form the mid-rib of the frond in those regions where the expanded wings or alae are developed, these latter being of varying width with either entire or serrate margins. Branching is generally dichotomous and the degree of branching increases as the rate of growth decreases.

In many species the branches bear swollen vesicles or *pneumatocysts.* In at least one species, *F. vesiculosus,* these are known to be seasonal in formation, but the nature of the habitat is also involved as more are produced in quiet than in rough waters. Their production appears to be linked with high photosynthetic activity in the spring and summer as the proportions of contained oxygen and nitrogen vary accordingly; carbon dioxide seems to be absent. In at least one species, *F. serratus,* the overall growth rate not only varies with season but also with the degree of shelter.

F. ceranoides will not grow in pure seawater and this species is therefore restricted to the brackish water of estuaries.

With increasing age the lower portions of the algae may be frayed off by wave action, leaving only the mid-rib, which then has the appearance of a stipe. The whole of the expanded thallus is covered with sterile pits or crypto-stomata similar to those of *Saccorhiza,* but in fruiting plants it is only the ends of the branches (receptacles) that become swollen and studded with the fertile conceptacles. In *F. spiralis* these conceptacles are hermaphrodite, containing both oogonia and antheridia; in *F. vesiculosus* and *F. serratus* the plants are dioecious, the two types of sex organs occurring on separate plants, while in F. *ceranoides* either state may be found. It has recently been shown that the auxin content of male receptacles of *Fucus* is higher than in female receptacles.

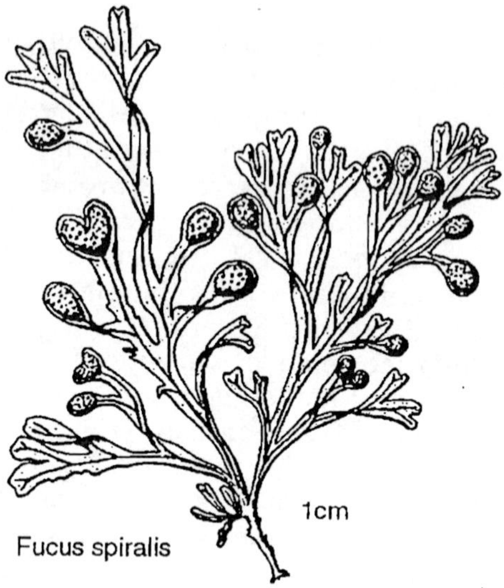

Fig. Fucus

In *F. vesiculosus* the reproduction peak occurs in spring and summer, but in *F. serratus* in autumn and winter. This may possibly be related, as in *Laminaria,* to differences in germination time and survival of plants, though no such evidence has been reported. The mortality rate of sporelings is extremely high in the early stages of growth (this also may possibly be seasonal), largely due to limpets. A number of very peculiar forms have been found which commonly occur on salt marshes: these rarely fruit, reproduction being secured by means of vegetative proliferations. Other forms suggestive of hybridization occur on rocky coasts, as well as probable hybrids between *F. vesiculosus* and the brackish-water species, *F. ceranoides.*

Recent work from Russia suggests that the maximum age of *Fucus* plants may be rather greater than that given above. In the Barents Sea, *F. vesiculosus* plants reach their maximum size at the greatest depth for the species (3 m.)

Morphologically the primary thallus is built up by the activity of the apical growing cell and by the surface layer of cells, the limiting layer or meristoderm. Below the limiting layer is a cortex composed of several layers of parenchymatous cells which become more and more elongate and mucilaginous towards the centre, and these probably form the storage system.

In the very centre the cells are extended into hyphae which are interwoven into a loose tangled web, but they do not extend right up to the apex. This central tissue (medulla) probably acts as a conducting system, because the transverse walls of the hyphae are frequently perforated with the same type of pit that is to be found in some of the Laminariaceae.

The primary medullary hyphae are relatively thin-walled, but when secondary growth of the thallus takes place the new hyphae which result are very thick-walled and are probably mechanical in function. Secondary growth is due to the activity of the limiting layer and the inner cells of the cortex, which also forms the secondary hyphae, and these penetrate between the primary medullary hyphae and finally outnumber them.

There is a greater development of secondary thickening in the stipe and mid-rib than there is in the frond, while in very old parts of the thallus the limiting layer may die off and then the underlying cortical cells take over its function. Development of secondary hyphae is most pronounced near the basal disc, which ultimately consists of a mass of hyphae.

Growth in length takes place by means of an apical cell which lies at the bottom of a slit-like depression that has resulted from the more rapid growth of the surrounding limiting layer. The apical cell is three-sided in young plants while in the adult thallus it becomes four-sided, the new segments being cut off successively from the base and four sides, after which they develop into the various tissues. Injury, and also the stimulus provided when the thallus lies on marsh soil, induces new growth in the neighbouring cells, and in this way proliferations are formed which may also serve for vegetative propagation.

Both cryptostomata and conceptacles arise as depressions in the surface of the thallus, and it is now known that they can arise in one of two ways:

1. A linear series of two or more cells is formed, but their horizontal activity then ceases, and the terminal initial cell becomes sunk in a depression as the

surrounding tissues grow up. The sides of the conceptacle are derived from the limiting layer and underlying cortex, e.g. *Fucus,* though in *Himan- thalia* only the limiting layer is involved. *Pelvetia canaliculata* forms an exception in that the first division is vertical and not horizontal. The floor of the conceptacle, however, is derived by division of the original basal cell of the linear series. Finally, around the remnants of the one or more initial cells, a central mucilaginous column is formed, stretching to the neck of the conceptacle and connected to the walls by thin strings of mucilage, which are later ruptured.

2. The cryptostomata or conceptacle develops from a single initial cell that divides transversely into two unequal cells; the upper or tongue cell degenerates while the lower one gives rise to the walls of the conceptacle by lateral divisions, except at the very top where they came from adjacent tissue. This method of formation occurs in *Sargassum, Bifurcaria.* The tongue cell and apical cell of the linear series are regarded as vestigial hairs so that the conceptacles really originate from the basal cell of a hair. The cryptostomata or hair pits may be regarded as a permanent juvenile stage of the fertile conceptacle.

In the mature fruiting conceptacles there are branched hairs or paraphyses with the antheridia borne terminally on the branches near the base; or else the paraphyses are unbranched and associated with the oogonia, which are either sessile or borne on a single stalk cell. Each oogonium characteristically contains eight ova when mature.

In those species where the conceptacles are hermaphrodite all these structures occur together. The expulsion of the gametes normally takes place at low tide, because the conceptacle is then full of mucilage and the loss of water makes the thallus shrink, so that the ripe ova and antherozoids in their envelopes are forced through the ostiole to the surface.

When the tide returns the inner wall bursts and so liberates the antherozoids. The antheridium wall is two-layered but the oogonium wall has three layers, the endochiton, mesochiton and exochiton. When the ova are ripe the exochiton ruptures and liberates the eggs through the ostiole enclosed in the meso- and endochiton. Next, the mesochiton ruptures and inverts itself, to expose the endochiton which is finally dissolved. The eggs are then set free into the water where fertilization takes place, the antherozoids clustering around the ova, until eventually one penetrates and fertilizes the ovum. A fertilization membrane is then produced.

The fertilized zygote divides at first by three transverse wall and becomes club-shaped. It is likely that light plays a part in determining the polarity of the zygote, one to two hours of weak light producing a detectable effect. Increasing the light intensity does not shorten the induction period. Other possible determinants of polarity are contact with a surface or point of entry of the antherozoid. The lowest cell elongates to form the first rhizoid while the uppermost cell divides twice by two longitudinal walls at right angles to give a quadrant. Further divisions result in the development of a central group of cells, the primary medulla, which is inside the cortex.

Additional rhizoids are produced and the apex of the embryo becomes flattened and one cell produces a hair with a basal intercalary meristem. Excessive growth of neighbouring cells results in the formation of a terminal depression and additional hairs are produced. Next all the cells of the first hair, except the basal one, disappear and the basal cell becomes the three-sided apical cell. In those Fucales, such as *Fucus*, where the adult condition of the apical cell is four-sided, this change in the apical cell takes place quite early, the new four-sided cell being cut off from the original three-sided one. The peculiar salt-marsh fucoids are interesting in that the juvenile three-sided apical cell persists through-out life.

In the allied genus *Pelvetia* the fronds have no mid-rib and are linear, compressed or cylindrical with irregular dichotomous branching. Air vesicles may be present in some species but ordinarily are absent. The structure of the thallus is essentially similar to that of *Fucus*, but the Californian *Pelvetia fastigiata* also possesses a few cryptostomata which are otherwise absent from the genus.

Only two ova mature in the oogonia, the remaining six

nuclei being extruded from the cytoplasm, though in *Pelvetia fastigiata* there may occasionally be four ripe ova or else ova that contain two nuclei. In *P. canaliculata* the two mature eggs are arranged one above the other, while in the Japanese species, *P. wrightii,* they are placed side by side.

The thallus of the common *Ascophyllum nodosum,* which sometimes bears nodular galls caused by the eel-worm *Tylenchus fucicola,* is more or less perennial, and regenerates each year from a persistent base or from the denuded branches. As in *Fucus* and *Pelvetia,* free-living or embedded forms have evolved in salt-marsh areas, and these differ not only vegetatively but also in the absence of reproductive organs. The normal fronds have a serrated margin but no mid-rib and commonly bear vesicles *(pneumatocysts).* When the vesicles are borne on the little side branches they are termed pneumatophores.

Recent work has shown that in some regions at least one pneumatocyst and its associated axis is produced annually so the number of pneumatocysts may indicate the age of the plant. This, however, does not seem to be universal. Work in Norway has shown that the pneumatocysts are only formed at a certain time of year, which ranges from the end of January in Eire to May in Canada, and formation is therefore probably related to sea temperatures.

The axis is beset by simple, clavate, compressed branchlets that arise singly or in groups in the axils of the serrations. These are later converted into or are replaced by shortstalked, yellow, fertile branches which fall off after the gametes have been liberated from their conceptacles. The oogonia each give rise to four ova and the remaining four nuclei degenerate.

HIMANTHALIACEAE: HIMANTHALIA

The short, perennial frond or button arises from a small disc-like holdfast; the shape of the button depends on its level, the frond being short and stumpy when it grows exposed at high levels, and more elongate at the lower levels where the plants are submerged for longer periods. From March to July of each year new receptacles grow out from the centre of the buttons and form very long strap-shaped and repeatedly-forked structures filled with mucus.

Although the button is regarded as the frond and the thong as the receptacle, it is possible to regard the button as part of the stipe and the thong as frond plus receptacle, as in *Durvillea.* In support of the second interpretation it should be mentioned that the thong contains horizontally running hyphae which have so far only been found in *Durvillea* in receptacular regions. In the thong these horizontal hyphae appear to be associated with the formation of air spaces.

Fig. Cystoseira

Growth curves show that the greatest length is attained by these annual thongs on plants growing in the lowest part of a dense zone, and that the shortest occur in the highest. This may be correlated with *(a)* the greater degree of desiccation at the higher levels, and *(b)* less frequent flooding which could reduce the supply of nutrient salts. The apical groove, containing the apical cell and hairs, is perpendicular to the plane of flattening.

Internally it has been found that the extremely elongated cells of the medulla may take on an appearance resembling the trumpet hyphae of *Laminaria.* This may be a case of parallel development, but in view of other similarities to *Durvillea* it may be suggested that *Himanthalia* represents a primitive though somewhat specialized member of the Fucales. Reduction has gone so far in this genus that only one ovum matures in the ripe oogonium (macrosporangium). Liberation of gametes is controlled by the tides and exposure, and there is a definite periodicity related to these two factors.

CYSTOSEIRACEAE: CYSTOSEIRA

The much-branched perennial thallus is either cylindrical or compressed and arises from a fibrous woody holdfast which has more or less the structure of a conical cavern. The primary branches arise from the main stipe towards the base and divide above into filiform branches and branchlets, which sometimes do not develop very far. Seriate rows of small air vesicles are present on small branchlets. In the related *Halidrys* similar rows are also present, but these are distinctly flattened and borne in leaf-like appendages. Just before the dormant season some species shed a considerable proportion of the branch system so that the plant may appear very different.

The plants are monoecious or dioecious, the conceptacles being borne in terminal or intercalary positions on the ramuli; as in some of the other genera, only one ovum develops and the remaining seven nuclei degenerate. In the sporeling the main shoot is very short and soon stops growth, and a new apical cell arises near its base.

This new apical cell gives rise to the adult axis which is therefore a branch of the sporeling. The later laterals develop monopodially and in succession. The first two shoots arise opposite each other but the remainder has a divergence of two-fifths. The genus is principally confined to the warm sub-tropical and temperate waters; in the Mediterranean considerable speciation and hybridization has taken place.

SARGASSACEAE: SARGASSUM

In the simplest forms branching is distichous but in the great majority it is radial with a divergence of two-fifths. Like *Cystoseira* the main axis is usually very short and the thallus is composed mainly of richly branched, long laterals which produce leafy, short laterals that are built up as follows: the primary branch is a sterile phylloclade which bears cryptostomata, while the secondary branch is also sterile and is commonly reduced to an air bladder.

In the simpler forms the subsequent branches are fertile and finger-like in appearance. In more advanced forms there may be further sterile phylloclades which may subtend axillary branch systems. In branching the main apical cell cuts off a secondary initial and this latter, as soon as it has emerged from the apical pit, cuts off another initial, on the distal side.

This last initial divides more rapidly than the other and grows out into the subtending leaf, which is therefore a lateral that has pushed aside the parent axis. Later the parent initial produces further initials on a two-fifth divergence. The plants are attached by means of a more or less irregular, warty, solid, parenchymatous base or else numerous stolon-like structures grow from the main axis and anchor the plant.

The genus, which is principally confined to tropical waters, is a very large one with about 150 species, some being dioecious although others are monoecious. Morphologically the structure of the species is similar to that of *Fucus* though there is little or no medullary material, and when present it is not gelatinized. In the ripe oogonium (megasporangium) only one ovum will normally reach maturity, though occasionally eight eggs may develop. In the former case the single ovum contains all eight nuclei, but only one of these grows larger and is actually fertilized.

This can be interpreted as a failure on the part of the megaspores and gametophytes to form cell walls, and is a secondary condition due to still further reduction. In *S. filipendula* there is no stalk to the young megasporangium and so it is embedded in the wall of the conceptacle. When ripe the whole oogonium, not merely the inner wall and its contents, is discharged and remains just outside the ostiole attached to the conceptacle wall by a long mucilaginous stalk which is a secondary development. In one sub-section of the genus the ostiole of the female conceptacle is closed by a disc-shaped gelatinous plug, which disappears when the oogonia are ripe.

After fertilization the first divisions take place while the zygote is still attached to the parent plant by this long stalk. The genus is very well known because of *S. natans*, the so-called Sargasso weed, which is found as large floating masses in the Sargasso Sea. At one time it was thought that plants of *S. natans*, together with one or two other pelagic species, were attached in the early stages, but there would now seem to be good evidence that they remain floating throughout the whole of their life cycle. It has been suggested that these perennial pelagic species originally arose from attached forms such as *S. filipendula* and *S. hystrix*. Pelagic species do not appear to possess the normal sexual reproduction but reproduce by vegetative fragmentation.

HORMOSIRACEAE: HORMOSIRA

The thallus, which looks like a bead necklace, is composed of a chain of swollen vesicles (internodes) with narrow connectives (nodes). Growth takes place by means of four apical cells, which give off branches alternately in a dichotomous manner, the branches usually arising at the internodes. Apart from the discoid holdfast, there is no differentiation into appendages comparable to those of the other Fucaceae.

The basal internode is solid but all the remainder are hollow; the nodes are also solid because they are composed solely of epidermis and cortex. The plants are dioecious, the conceptacles being borne on the periphery of the inflated nodes. Although eight ova are originally formed in the oogonium only four reach maturity, but in this genus, however, it is a case of degeneration of eggs and not merely of nuclei. When the ova are released from the oogonium they are still surrounded by the endochiton.

This soon ruptures and releases the mature ova. After fertilization has taken place a special fertilization membrane is produced, and this is composed of two layers, the inner one being of cellulose. Another interesting feature of this genus is its capacity to form and shed a cuticle that bears the impressions of the cell outlines. The genus is monotypic, the single species, *H. banksii*, being confined to Australia and New Zealand where it grows on rocks and in tide pools of the littoral belt.

Although several varieties have been described, it has recently been shown that these are merely more-distinct representatives of a range of ecological forms. Use of a statistical technique, the discriminant function, based upon measurements of bladder diameter and connective length, has differentiated these various ecological forms.

6
Rhodophyta

BANGIACEAE: PORPHYRA

This genus has a wide distribution extending from 15° to 71° N. and from the Cape of Good Hope to 60° S. The plant is flat and membranous, and in the common European species, *P. umbilicalis,* there are a number of growth forms; the shape, width and length of the various forms are determined by the age of the plant, the height above sea-level and the type of locality.

The plants are attached by means of a minute disc which is able to produce lateral extensions from which new fronds may be proliferated. The disc is composed of long slender filaments together with some short stout ones, those near to or in actual contact with the substrate swelling up, branching and producing suckers or haptera which can penetrate dead wood or the tissue of brown fucoids. In the latter case there is evidently a capacity for epiphytism once contact is secured, and there is even some evidence of partial parasitism.

The gelatinous monostromatic fronds of *Porphyra,* which become dis-tromatic during reproduction, are composed of cells that possess one or two stellate chloroplasts with a pyrenoid, the process of nuclear division being intermediate between mitosis and amitosis. Reproduction is by means of mpnosporfis, carpogonia with rudimentary trichogyngsjind antheridia. The carpogonial areas occupy a marginal position on the thallus and because of pigment differences they exhibit a different colouring to the vegetative thallus.

In the Japanese *P. tenera,* monospores are shed in considerable numbers, 10,000 spores being produced from 100 mm^2 of thallus. These become attached to the substrate in a short space of time and commence germination to a new plant one to two days after fertilization. In sexual reproduction all the frond, except the basal region, can produce antheridia.

Fig. Porphyra.

In the case *of P. tenera* of Japan, penetration of shells takes place within three days. This dwarf phase, which cannot tolerate any real degree of exposure, bears sporangia that produce monospores, and these on germination are thought to give rise to a new adult plant, though some workers consider that they only reduplicate the "*Conchocelis*" phase. New adult plants of P. *umbilicalis* and *P. perforata* are said to arise from spores (conchospores) liberated from short swollen branches (fertile cell rows) or else arise as vegetative buds.

Japanese workers consistently report monospores from the '*Conchocelis*' phase of P. *tenera,* whereas in the case of P. *umbilicalis* only conchospores and buds are reported.

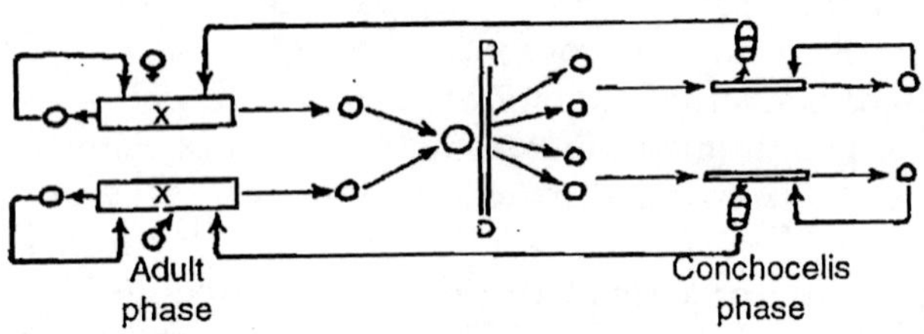

Fig. Diagram of Life Cycle of Porphyra Umbilicalis.

The development and growth of both adult and '*Conchocelis*' phase is related to day length and this also varies for the species. In *P. tenera* 2-12 hours of light is all that is required for production of monosporangia and liberation of spores. Short days are required for maturation of '*Conchocelis*[9] monosporangia in *P. yezoensis, P. angusta,* and *P. kuizieda,* while very short days are required *forP.pseudolinearis.*

Production of the '*Conchocelis*' phase is not universal because in New Zealand, plants from the north produce '*Conchocelis*' but similar plants from the south apparently do not. Although the general pattern cf the *Porphyra* life history is known, many specific cases still need to be worked out. A *Conchocelis* phase has also been reported recently for the allied *Smithora naiadum*.

Haploid chromosome numbers of two *(P. umbilicalis)*, three *(P. tenera, P. yezoensis)*, four *(P. linearis)* and five have been listed. In *P. umbilicalis* and its variety the same chromosome number is reported from both adult and '*Conchocelis*' phase. In the light of what is now known about red-algal life histories one would expect the dwarf phase to be the diploid generation in which case reduction division should be observed in the monosporangia or fertile cell rows.

The plants consist of branched filaments, or are compact pseudoparenchy-matous or membranous thalli with uniaxial or multiaxial type of construction. Fundamentally they are heterotrichous though this condition is lacking in many. Growth is normally from one or more apical cells and plasmodesmata are present.

The carpogonium is usually highly specialized and only one spermatium is produced from each'spermatangium. The zygote nucleus commonly passes into a special auxiliary cell and after fertilization there is a parasitic phase, the carposporophyte, producing carpospores. This is nearly always diploid and the carpospores give rise to an asexual generation by means of tetraspores.

ACROCHAETIACEAE: ACROCHAETIUM

The genera of this family are characterized by a uniaxial uncorticated construction of branched filaments. Many of the species are minute epiphytes or endophytes with the monosiphonous branched threads creeping between the host cells. *Acrochaetium* is the largest genus and of world-wide distribution.

In certain species there is often a basal layer that gives rise to erect monosiphonous branched threads in which each cell contains one chloroplast so that the plants are heterotrichous. Lateral branches sometimes terminate in a long cellular hair. Reproduction is primarily by monospores (occasionally bispores or tetraspores) produced on lateral or terminal sporangia.

Recently it has been shown that an *Acrochaetium-like* alga is the alternate generation of *Liagora* and many species of *Acrochaetium* for which only asexual reproduction is known may in fact be alternate generations of other Rhodophyceae. Sexual reproduction is known for a few species which may be monoecious or dioecious. Groups of two or three spermatangia are formed on small side branches and carpogonia are also formed.

After fertilization the gonimoblasts grow out from the base of the car-pogonium, the terminal cell of each row becoming a carposporangium. In allied genera alternation of the sexual and asexual generation has been reported. In *Rhodochorton floridulwn* the macroscopic plant reproduces by tetraspores which give rise to a dwarf generation that may be haploid. In *R. purpureum*, West has successfully grown the slightly smaller game-tophyte generation in culture. Further work on members of the *Acrochaetium* assemblage is needed in order to establish the general character of the life histories.

BATRACHOSPERMACEAE: BATRACHOSPERMUM

B. moniliforme, which is a very variable species, is found attached to stones in swift-flowing fresh waters of the tropics and temperate regions. The thallus, which is violet or blue-green in colour, is soft, thick and gelatinous, and the primary axis, which grows from an apical cell, comprises a row of large cells. Numerous branches of limited growth arise in whorls from the nodes and often terminate in long hairs. The basal cell of such a branch can give rise to a branch of unlimited growth or else to corticating cells that grow down and invest the main axis.

All cells are uninucleate and contain one chloroplast and pyrenoid. Reproduction takes place by means of monospores (restricted to the juvenile phase), carpogonia and antheridia. The antheridia arise as small, round, colourless cells at the apices of short, clustered, lateral branches. The carpogonia are also terminal and possess a trichogyne separated by a constriction from the rest of the organ; this shrivels after fertilization.

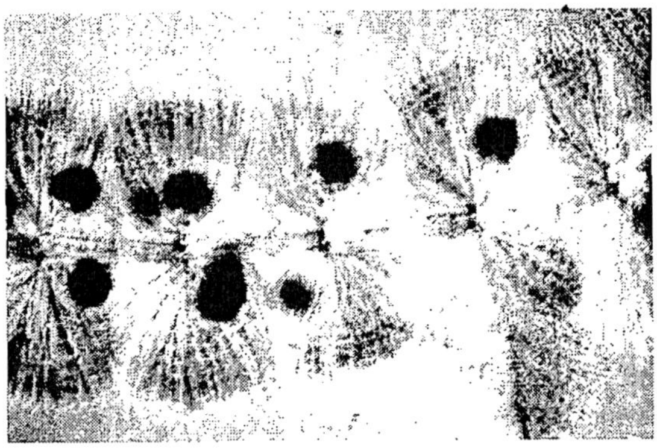

Fig. Batrachospermum Moniliforme

The zygote nucleus divides twice giving rise to four nuclei; these pass into protuberances that form from the carpogonium and then branch to give a mass of gonimoblast filaments. Each branch terminates in a sporangium that produces a single naked carpospore. On germination it gives rise to a diploid microscopic, branched filamentous stage, which, on account of earlier confusions, is now known as the *'Chantransid'* stage.

Practically all freshwater species of *Chantransia* represent a phase in the life cycle of a species of *Batrachospermum.*

The *'Chantransia'* phase can reproduce itself indefinitely, but eventually an apical cell undergoes reduction division, three nuclei abort and the new haploid cell gives rise to a new adult *Batrachospermum* plant. The gametophyte is therefore parasitic on the tetrasporophyte while the carposporophyte is parasitic on the female gametophyte.

The freshwater *Lemanea* (Lemaneaceae) has a more compact, pseudo-parenchymatous thallus and much the same kind of structure is found in the marine Naccariaceae. In *Lemanea rigida,* Magne has shown that the carpospores have double the number of chromosomes found in the vegetative cells. The nature of the sporophyte is at present unknown but one may expect it to be a dwarf filamentous alga by analogy with other Nemalibnales.

In L. *catenata* the carpospore germinates into a diploid protonema from which the adult arises after reduction division in the apical cell of a cladome. The peculiar freshwater genus

Thorea (Thoreaceae) reproduces only by monospores and appears to have lost sex organs.

HELMINTHOCLADIACEAE: LIAGORA

The axial portion of the 'fountain'-type thallus contains large cells with which are intermingled narrower rhizoidal cells. Calcium carbonate is sometimes deposited to a greater or lesser extent on the outside of the thallus so that the plants appear to be white. One important feature is that in four species the carpospores divide to give four spores that have been regarded as tetraspores.

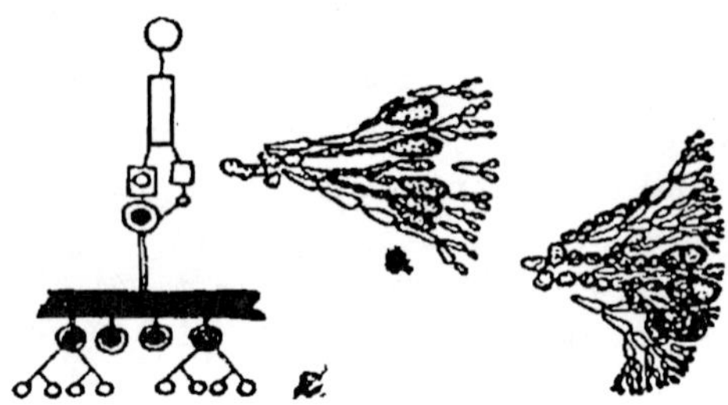

Fig. Liagora

Although no cytological evidence is available it is presumed that meiosis is delayed to the time when the carpospores divide. There is thus no independent tetrasporic diploid generation. This phenomenon was first discovered in the Atlantic L. *tetrasporifera.* A similar state of affairs has more recently been recorded for other species of *Liagora* and for *Helminthocladia agardhiana.*

This condition is not universal within the genus *Liagora* because recently it has been shown that the carpospores of *L. farinosa* germinate to an *'Aero-chaetium'-like* plant that produces, after meiosis, tetraspores that yield a new *Liagora* plant. The dwarf phase can be perpetuated by means of monospores. In this species, therefore, there is an alternation of unlike generations, as there is in *Nemalion helminthoides* and in *N.pulvinatum.*

BONNEMAISONIACEAE: ASPARAGOPSIS

The erect shoots arise from a perennial creeping thallus

which possesses vesicular or iodine-containing cells whilst modified branches act as attachment organs. In the erect thallus primary and secondary laterals form a cortex which is often separated from the original axis by a space. In all Bonnemaisoniaceae except *Asparagopsis*, the thallus is flattened and the appendages arranged in two rows.

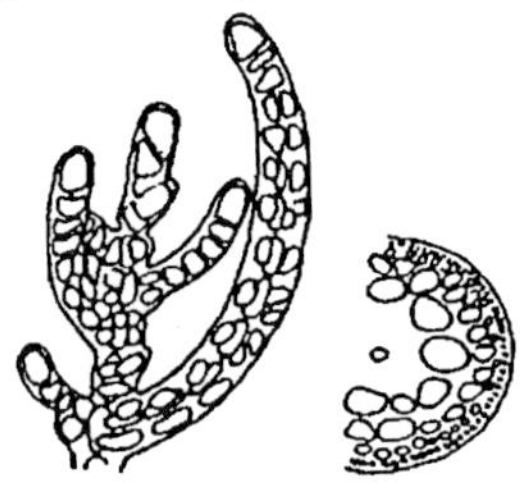

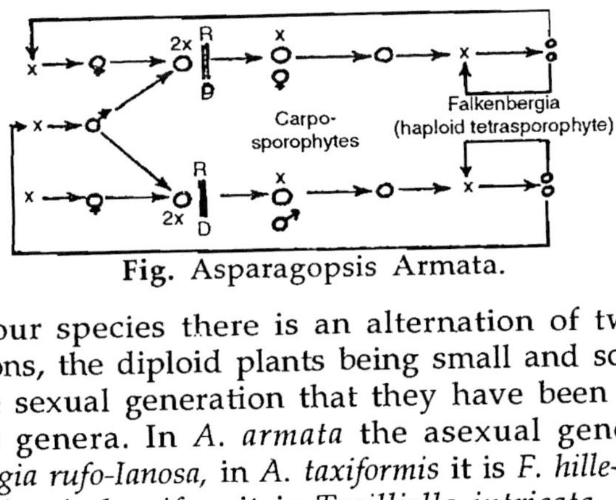

Fig. Asparagopsis Armata.

In four species there is an alternation of two unlike generations, the diploid plants being small and so different from the sexual generation that they have been placed in separate genera. In *A. armata* the asexual generation is *Falkenbergia rufo-lanosa,* in *A. taxiformis* it is *F. hille-brandii,* in *Bonnemaisonia hamifera* it is *Trailliella intricata,* and in *B. asparagoides* it is *Hymenoclonium serpens.* In the last named species it is possible that meiosis occurs in the carposporangia because the dwarf *Hymenocloniwn* plants give rise vegetatively to the adult gametophytes.

In the genera *Leptophyllis* and *Delisea* isomorphic tetrasporic generations have been recorded for *Leptophyllis conferta, D. suhrii* and also for *D.fimbriata (D.ptilchrd)* in Australian waters, though the last species is reported as having only the haploid generation in Japanese waters. The same kind of haplophasic life history is also reported for the Japanese species *D. elegans, D. hypneoides* and *Ptilonia okadai.*

There are therefore three types of life history to be seen in the family:

1. Asparagopsis type with a dwarf asexual generation, which can be either diploid (A. armatd) or haploid (B. asparagoides).
2. Ptilonia okadai type with only the sexual generation and carposporophyte.
3. Polysiphonia type with isomorphic alternation.

CHAETANGIACEAE: SCINAIA

The reasons for uniting the Chaetangiaceae with the Bonnemaisoniaceae have been outlined by Fan. They are based primarily on the function of the hypogynous cell of the carpogonial branch as a nutritive auxiliary cell similar to that *of Asparagopsis.*

Scinaia is a widespread genus with most of its species in the northern hemisphere. The corresponding genera in the southern hemisphere are *Pseudoscinaia* and *Pseudogloiophloea.* The fronds, which arise from a discoid holdfast, are subgelatinous, cylindrical or compressed and dichotomously branched. The centre of the thallus is composed of both coarse and fine colourless filaments, the former arising from the apical cell and the latter from the corticating threads. There is also a peripheral zone of horizontal filaments that terminate in short corymbs of narrow assimilatory hairs interspersed with large colourless cells.

These two types of epidermal cells are apparently differentiated near the apex of the thallus, the small ones giving rise to hairs, monosporangia or antheridia. The large colourless cell is thought to give protection against intense light, but it may also be a relic of a tissue which formerly had a function that has since been lost. One of two spores are formed in each monosporangium, while the spermatia arise in sori, forming bunches of cells at the ends of the small-celled assimilatory branches.

The commonest species, *S. furcellata,* is monoecious, and the plants bear three-celled carpogonial branches, the reproductive cell containing two nuclei, one in the carpogonium proper and one in the trichogyne. The second cell of the carpogonial branch gives rise to a group of four large cells which are rich in protoplasm, while the sterile envelope of the cystocarp arises from the third cell.

After fertilization the zygote nucleus passes into one of the four large cells. From this large cell a single gonimoblast filament grows out which later becomes branched. Recently, the carpospores have been shown to give rise to a filamentous gametophyte. This represents another example of a heteromorphic alternation of generations.

In the allied genus *Galaxaura* the diploid carpospores give rise to a tetraspore generation differing morphologically in some degree from the sexual generation. There is thus an alternation of unlike generations, though the difference is in appearance rather than in size.

GELIDIACEAE: GELIDIUM

In the stiff, cartilaginous thallus, which is often ginnatelv branched, the uniaxial construction is most clearly seen at the apex. But in the older thallus there is a central portion of large cells, derived from four primary pericentral cells, with a peripheral zone of small cells with interspersed hyphae. The presence of special nutrient cells in the vicinity of the carpogonial branch results in the production of a complex structure composed of several carpo-gonia with small nutrient cells in short branches and only one of these carpogonia needs to be fertilized.

After fertilization a single gonimoblast filament initially grows out into the mass of nutritive cells, but it later branches so that a number of terminal carposporangia are finally formed. In *Gelidium* the cystocarp wall develops two pores or ostioles, one on either side of the supporting branch has only one ostiold.

The carpospores germinate to give a tetrasporophyte morphologically identical to the gameto-phyte.fjln the Japanese *G. amansii*, the principal source of agar, the tetraspores are shed when the water temperature rises to 20° C and the carpospores when it rises to 24° C, and this shedding occurs each afternoo. The spores become fixed to the rock surface when they have been in contact for about ten minutes. The maximum temperature for germination is about 25°-26° C.

The sporelings can grow in brackish water and this

markedly affects the length of the primary rhizoids. Growth is slow at first but later becomes very rapid. More tfun one erect shoot may arise from each basal pad or the lateral branch initials may grow horizontally giving rise to stolons from which more initials may arise.

The members of this order are very diverse in habit. A number are uniaxial and similar in structure to the Gelidiales, while others are multiaxial. There is a whole group in which there is extensive lime encrustation and considerable reduction of the thallus, e.g. *Lithothamnion.* Definite auxiliary cells are produced on accessory branches prior to fertilization and these serve not only for nutrition but also as starting points for the gonimoblast filaments.

The order divides naturally into two main subdivisions. In the first group the primary gonimoblasts fuse with nutritive cells in the carpogonial branch and secondary gonimoblasts then fuse with the auxiliary cells. In the second group the primary gonimoblasts fuse directly with auxiliary cells. In the first group are the Dumontiaceae and here the carpogonial branch and the auxiliary cell branch are widely separated.

In the remaining families of this group the carpogonial branches are borne in nemathecial groups. The second group is also subdivided according to whether the carpogonia and auxiliary cells are borne on separate branches as in the Grateloupiaceae or on an adjacent branch. In certain genera, e.g. *Callophyllis*, *Weeksia*, procarps are present. There are twelve families in the order, but some of these are small and unimportant.

DUMONTIACEAE: DUDRESNAYA

The cylindrical, highly branched thallus arises from a prostrate disc. It is soft and gelatinous and consists when young of a simple, articulated, filamentous axis with whorls of dichotomously branched ramuli of limited growth. In older plants the axis becomes polysiphonous and densely beset with whorls of branches. The polysiphonous condition is produced by enlargement of the basal cells of the primary branch whorls and of cortical threads produced from them.

The plants are dioecious, the males being somewhat smaller and paler. The carpogonial branches of *D. coccinea* arise from lower cells of short side branches and when fully developed consist of seven to nine cells; they branch once or twice and may have short sterile side branches arising from the lowest cell.

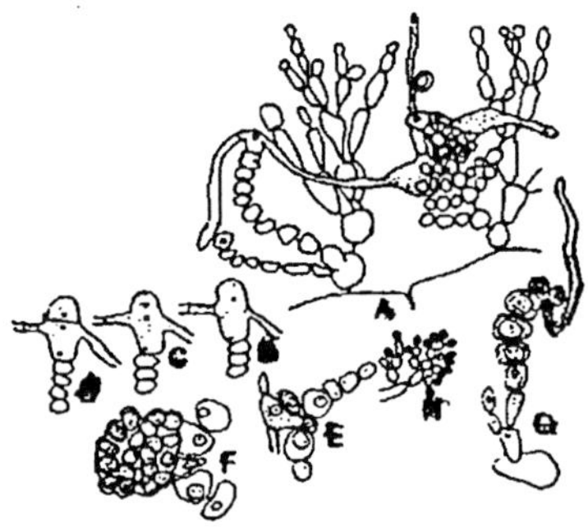

Fig. Dudresnaya

In the middle of the mature carpogonial branch there are two to three larger cells which function in a purely nutritive capacity, while the auxiliary cells develop in similar positions on neighbouring homologous branches. After fertilization the carpogonium sends down a protuberance containing the diploid nucleus and this cuts off two cells when it is near to the nutrient cells of the carpogonial branch. These all fuse together and sporogenous threads, each carrying a diploid nucleus, then grow out towards the auxiliary cells on the other branches.

When these filaments, or connecting threads, fuse cytoplasmically with an auxiliary cell, the latter forms a protuberance into which the diploid nucleus of the connecting thread passes. When the nucleus has divided the protuberance containing one of the daughter nuclei is cut off by a wall. The gonimoblast filaments then grow out as a branched mass from this protuberance. Each sporogenous thread sent out from the original cell may unite with more than one auxiliary cell in the course of its wanderings through the thallus, so that one fertilization may result in the production of a number of carposporophytes.

SQUAMARIACEAE: PEYSSONELIA

The characteristic members of this family are crusts of which *Peyssonelia* is the most common. The thallus is blood red and more or less firmly attached to the substrate; in some species there is lime encrustation. It commences as a monostromatic layer that gives rise on the upper surface to sparingly branched threads that become fused to each other to give a parenchymatous-like tissue. Rhizoids develop on the lower side of the primary thallus and serve as attachment organs. The spermatangia are borne laterally on erect threads that are in soral groups or nemathecia.

Pericentral cells are cut off and these then act as spermatangial mother cells. The carpogonia are also borne in nemathecia. Within the nemathecium there are six to eight celled paraphyses, the lowermost cell of which can give rise to a carpogonial branch or an auxiliary cell branch. Both of these branches contain four cells. After fertilization the carpogonium fuses with the second cell of the carpogonial branch and then from this fusion cell a primary gonimoblast thread grows out to fuse with the second cell (auxiliary cell) of an auxiliary-cell branch.

The secondary gonimoblast is then formed. It comprises 8-12 cells all of which, except the basal, become carposporangia. In the tetrasporangiate plants the cruciate sporangia are also borne in nemathecia.

HILDENBRANDIACEAE: HILDENBRANDIA

This genus also forms thin brownish-red crusts which adhere strongly to stones or other algae, and it is frequently difficult to distinguish from similar encrusting types such as *Peyssonelia.* The frond is horizontally expanded into a thin encrustation of several layers of cells arranged in vertical rows. The genus is both marine and fresh water, *H. rivularis* appearing frequently in rivers and streams.

Fig. Hildenbrandia Prototypus

The principal mode of reproduction is by means of fragmentation, stolons or tetraspores produced in sporangia borne in rounded or oval conceptacles that are sunk in the thallus. The absence of sexual plants suggests specialization, though it is possible that a protonemal gametophyte exists though this has not yet been observed.

CORALLINACEAE: LITHOTHAMNION;

Heavy calcification (calcite) is the^ftarSctenstic of this family which may be divided into two subfamilies:

1. The Melobesioideae, which comprises the crusts and nodular Corallina-ceae, e.g. *Lithothamnion, Melobesia, Heteroderma,*
2. The Corallinoideae, which is composed of the jointed Corallinaceae, e.g. *Corallina.Jania, Amphiroa, Bossea.*

In *Lithothamnion* there is a multi-layered basatzone of horizontally aligned cells, the hypothallium, and a surface layer of vertically aligned cells, the perithallium. The various reproductive organs are borne in conceptacles on separate plants. The aggregation of the reproductive organs into conceptacles must be regarded as equivalent to nemathecia which subsequently become overgrown by surrounding tissue. In the male plants there are a number of two-celled filaments in the centre of each conceptacle.

The basal cells of these threads cut off two elongate, antheridial mother cells, which in their turn produce two antheridia. Antheridial mother cells are also produced from the lower parts of the conceptacle walls. Since the antheridium is liberated entire it must be regarded as morphologically equivalent to a spermatium.

In the female plant the central threads of a conceptacle form three-celled carpogonial branches, the basal cell of the branch being an auxiliary cell. The outer threads of a conceptacle produce sterile two-celled filaments. Only the central carpogonia mature and usually only one of these is fertilized and produces carpospores. After fertilization the carpogonium and the cell below it fuse together and send out a filament to the lowest cell (auxiliary cell). Later all the auxiliary and nutritive sterile cells fuse to give one long fusion cell from which very short gonimoblast filaments grow out.

In the tetrasporic plant there are simple filaments which give rise not only to the tetrasporangia but also to branched sterile filaments that grow out and form the conceptacle roof by division and elongation. Finally the membrane of the tetrasporangia penetrates this roof and form exit pores. The spores are arranged in a linear series.

The multiaxial type of construction can be observed extremely well in *Corallina*. The erect plants, which are jointed, cylindrical or compressed, arise from calcified, encrusting basal discs or prostrate interlaced filaments. Branching is generally pinnate, but in the closely allied genus *Jania* it is dichotomous. There is a central core of dichotomously branched filaments with oblique filaments growing out at the swollen internodes to form a cortical layer, the end cells being flattened and the whole encrusted by lime except at the joints. The plants are monoecious or dioecious, the reproductive organs being borne in terminal or lateral conceptacles.

The carpogonia arise from a kind of prismatic disc formed from the terminal cells, which also function later as auxiliary cell. As a result of oblique divisions, one to three carpogonial branches are formed on each mother cell, but only one of these finally develops into the mature, two-celled carpogonial branch. After fertilization a long or rounded fusion cell is formed by the auxiliary cells, and this contains both fertilized and unfertilized carpogonial nuclei.

Fig. Corallina Officinalis

The antheridia are elongated, and after liberation the spermatia round off and remain attached to the antheridial wall by means of a long, thin pedicel in *C. officinalis* and by a short stalk in *Jania Rubens.* The tetrasporangia comprise a stalk cell and sporangium in which the four spores are arranged in a linear row.

GRATELOUPIACEAE: GRATELOUPIA

The plants vary in size depending on the species but they can be as long as 75 cm, with a narrow or broadly flattened axis, dichotomously or pinnately branched. Some species bear proliferations from the margin of branches or from the surface. The thallus characteristically feels firm and gelatinous. In section there is a central region with small, elongated, anastomosing filaments which arise from star-shaped cells, while the cortex consists of rows of dichotomously branching filaments of moniliform appearance.

The structure is therefore multiaxial. Primary gonimoblasts grow directly from the fertilized carpogonia to auxiliary cells borne on separate branches. The first cell of the secondary gonimoblast that arises from the auxiliary cell gives rise to numerous branches; nearly all the cells in the branches produce carpospores. The tetrasporangia are distributed throughout the cortical region in the sporophytic plants. There are a large number of species which are mostly restricted to the warmer seas.

KALLYMENIACEAE: CALLOPHYLLIS

This family differs from the preceding one in that the auxiliary cell is borne at the base of the carpogonial branch; it is in fact the support cell and with the carpogonium forms a procarp, an exceptional feature in this family. There are therefore no secondary gonimoblasts.

The membranaceous thallus is flattened with either dichotomous or palmate branching. The central region is occupied by a mass of large cells with smaller ones intermingled, while the cortex consists of rows of short filaments arranged at right angles to the surface. The cystocarps are normally borne in outgrowths from the margin of the frond.

CHOREOCOLACACEAE: *HARVEYELLA*

This and the closely allied genus *Holmsella* are monotypic

genera each containing a holo-parasitic species, and *Choreocolax* is another parasitic genus very nearly related to them. *Harveyella mirabilis* is parasitic on species of *Rhodomela* while *Holmsella pachyderma* is a parasite of *Gracilaria confer-voides.* They have little or no colour, as might be expected from parasites, and they send out branched filaments or haustoria into the tissues of the host. These penetrate the middle lamellae of the host cells and form secondary pit connections with them.

The parasites appear as external cushions lying on the branches of the host, each cushion surrounded by an outer gelatinous coat and consisting of a central area that is four to five cells thick. In *Harveyella* the female cushions bear numerous procarps with four-celled carpogonial branches, each with two sterile branches. In *Holmsella* the carpogonial branch is two-celled.

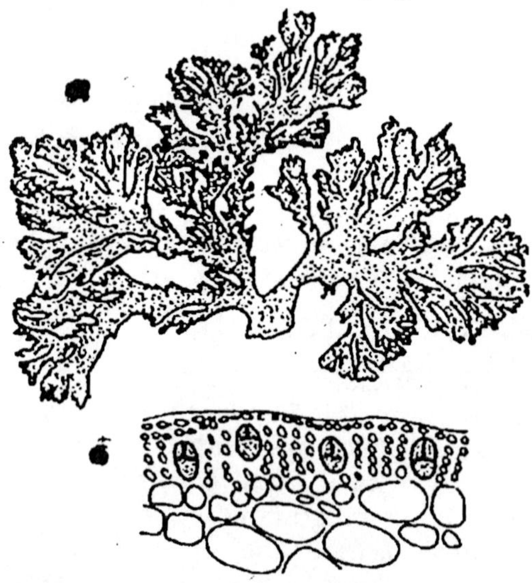

Fig. Callophyllis

The species are said to pass through the full, floridean life cycle twice every year. It is clear that their reduced morphological features are associated with their parasitic habit. In the past the family has been placed in the Gigartinales, but is now more generally placed in the Crypto-nemiales.

7
Cryptophyta Pyrrhophyta

CRYPTOMONADACEAE: CRYPTOMONAS

The species of this genus possess the typical naked cryptomonad cell. *Cryptomonas ovata* var. *palustris* has an olive-greenish hue whereas other genera may be red.

Apart from the usual longitudinal division of the motile cell, some species have thick-walled, resting-stage cysts. *Hemiselmis*, a representative of the Hemiselmidaceae, is characterized by a bean-shaped cell, with laterally inserted flagella. The gullet is replaced by an oblique furrow with trichocysts. There are one or two parietal chloro-plasts, which in *H. rufescens* are red but are blue-green in *H. virescens*.

Initially there is only a single chloroplast but during maturation this divides into two so that when division occurs the two chloroplasts distribute themselves between the two daughter cells.

CRYPTOCOCCACEAE: TETRAGONIDIUM

The one species in this order, *Tetragonidium verrucatum*, is a rarely found organism that has been recorded from freshwater ponds in Europe, the United States and New Zealand. The cells have a characteristic tetrahedral shape and a cellulose cell wall. Reproduction is by means of zoospores which possess the characteristic features of the flagellate Cryptomonads.

On coming to rest these zoospores develop into the typical coccoid parent. A new genus *Bjornbergiella* has recently been described. It exists in the form of filaments of a few cells each and together with *Tetragonidium* is the only indication of evolution towards an algal organization.

Like some other flagellated groups, the dinoflagellates

possess trichocysts. These consist of long membrane-bound crystalline cores connected by fibrils to their internal retainer sac. The trichocysts, which have a banded appearance, originate in vesicles of golgi bodies and are ejected through a fine pore at the cell surface. Their exact function is unknown but they are probably a mechanical sensing or defensive device.

The brownish colour of the chloroplast is due to the predominance of the group-specific carotenoid peridinin, which masks the chlorophyll *a* and *c*. Species without chloroplasts often possess cytoplasmic pigments, of unknown nature, which impart various colours to the cell. Some dinoflagellates are also characterized by the ability to luminesce and by the presence of potent neurotoxins. The storage reserve is starch.

The Desmophyceae and Peridiniales are predominantly holophytic (and nearly all are obligate photoauxotrophs), although holozoic nutrition is known to occur amongst phototrophic species. The Gymnodiniales are predominantly holozoic, as are the Dinophysiales and Rhizodiniales, while the Blastodiniaceae are parasitic. The non-flagellated orders are typically holophytic. In those dinoflagellates with holozoic nutrition ingestion takes place typically, but not always, while the alga is in a temporarily assumed amoeboid form.

Although the majority are free-swimming or attached, marine or freshwater forms, they are also quite common as sand dwellers and parasites in fish and invertebrates. The endosymbiotic algae (Zooxanthellae) of corals, sea anemones and certain other coelenterates and radiolarians are dinoflagellates. There are two classes within the division.

PROROCENTRACEAE: PROROCENTRUM

The cell shape varies from spherical to ovoid with noticeable flattening, except in some species of *Exuviaella*. The flagella project through the envelope via the individual pores at the edge of which there may be teeth-like projections.

One flagellum projects straight forward to provide the forward motion while the other, causing the rotatory motion, extends at right angles. There are usually two yellow-brown chloroplasts commonly with associated pyre-noids, and numerous trichocysts are often present. Reproduction is by longitudinal cell division, where one valve of the envelope is retained for each daughter cell, but cysts have been recorded in older cultures. *Prorocentrum* is a common marine genus and *P. micans* is often a component of red tides. *Exuviaella marina* is a sand dweller.

DINOPHYCEAE (DINOKONTAE)

This class represents the main line of dinoflagellate development, which is considered to be the more primitive, in certain important respects. The cell wall or envelope when present is not divided by a single vertical suture. The motile flagellates demonstrate the characteristic 'dinoflagellate orientation', because they are circled by a transverse or spiral groove, the girdle, which divides the outer wall into a forward portion, the epicone or epitheca and a hypocone or hypotheca (posterior part). Parallel to the axis of locomotion is a longitudinal furrow, the sulcus. Both the sulcus and girdle may be greatly modified, but the basic pattern remains the same.

The two flagella arise on the ventral side, usually through separate pores at the junction of the girdle and sulcus. The flagellum in the transverse groove is a helical ribbon possessing a unilateral row of long fine hairs and a single striated longitjjdJnaLsirand, while the longitudinal flagellum is pleuronematic with short fine lateral hairs, though these are absent in *Amphidinium*. Some workers have argued that the flagella are not latero-ventrally inserted, but still apically inserted and that the dinoflagellate motion is consequently a unidirectional sideways spiralling.

Apart from a few genera that have only a firm periplast, the Dinophyceae possess a cellulose wall. In a large majority of the flagellated genera, the 'armoured' dinoflagellates, the walls possess regularly arrange sculptured plates whose numHer and arrangement are taxonomically important. Those genera without sculptured plates or without the cellulose wall are referred to as 'unarmoured' dinoflagellates.

One or more variously sjiaped peripheral chloroplasts are present (except in colourless members) with or without pyrenoids, and showing the typical yellow-green golden-brown: colour, and^possess a triple outer membrane. In addition to photoauxotrophy both holozoic and^aprophytic nutrition are known, even in phototrophic species.

Asexual reproduction is by a cell division, or in the case of the non-flagellated genera, or *phytodinads,* by aplanospores or by *Gymnodinium-type* zoospores, a factor thaTls largely responsible for their inclusion in the Dinophjceae.

Sexual reproduction has been recorded for a few species and involves isogamy with amoeboid or motile gametes. The motile flagellates are divided into two orders, the Gymnodiniales (comprising the unarmoured Gymno-diniaceae and a few other small specialized families) and the Peridiniales (armoured dinoflagellates, comprising a large number of families). The unarmoured dinoflagellates are not well known because they are very difficult to preserve and need to be studied as they are collected.

GYMNODINIACEAE

Although the members lack a cell wall the pellicle is firm enough to give the cells a rigid shape which does vary greatly from one species to another. The cells are generally spherical although ventral or lateral flattening is quite common. From a ventral view the girdle spirals to the left, sometimes going more than once around the cell. The sulcus is usually straight but it too may be twisted around the cell. Cyst formation is quite common. The majority of species lack chloroplasts and are holozoic.

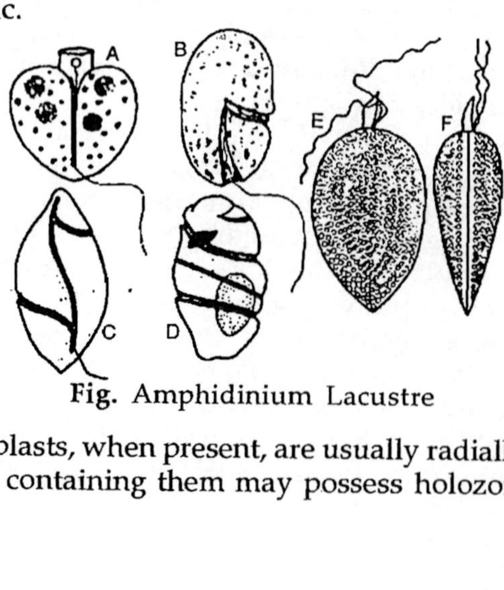

Fig. Amphidinium Lacustre

Chloroplasts, when present, are usually radially arranged, and species containing them may possess holozoic nutrition.

The two pusules that open to the two flagellar pores are sometimes fused into one. Certain species have been recorded as possessing functional trichocysts or cordocysts. The nucleus is usually centrally located. Cell multiplication is by longitudinal division of either the motile or non-motile encysted form.

In *Gymnodinium* the girdle is usually equatorial with little displacement of the two ends. In some species the periplast on the non-motile encysted cell may be striated.

In *Amphidiniwn* the girdle is anterioly placed so that the epicone is very small. The single chloroplast is arranged peripherally with a large external pyrenoid. The other cell components such as trichocysts, nucleus, pusules, flagella and so on, are typical of dinoflagellates although the trailing flagellum lacks lateral hairs. Many of the species are flattened, either laterally or dorso-ventrally, and are found predominantly in coastal waters. Isogamous sexual reproduction has been recorded for *A. carterae.* After fusion of two individuals, the zygote divides in one plane. The two subsequent divisions are unusual in that they are asynchronous.

In *Hemidinium,* the girdle is incomplete, extending usually about halfway round the cell, and the sulcus is restricted to the hypotheca.

In *Gyrodinium,* many species of which show cell elongation, the ends of the girdle are widely separated giving it a spiral appearance, while the sulcus may also be slightly twisted. Most of the species are colourless.

The spiralled appearance of *Cochlodinium* is due to the distinct girdle and furrow. The three following, widely distributed, colourless dinoflagellates possess an interesting morphology and are accordingly classified in separate families:

Polykrikos is one of the few coeno-cytic forms, the cell comprising two, four or eight transverse grooves with one, two or four nuclei respectively, because only one nucleus is present for each pair of grooves in the coenocyte. In each zooid the girdle is slightly spiral. The sulcus runs the entire length of the zooid and is continuous with that of each adjacent zooid, since each zooid possesses the same orientation.

Polykrikos is also characterized by the possession of complex functional barbs or nematocysts as well as trichocysts.

Oxyrrhis. This genus is considered to be a primitive gymnodinioid member. The transverse furrow and sulcus are both situated at the posterior end of the asymmetrical cell. The latter is rather broad and divided by a lobe-like structure. The species are colourless, holozoic and also phagotrophic. The cells of *Noctiluca miliaris,* the sole species of the genus, are very large, sometimes reaching a diameter of 2 mm. The girdle is non-existent and the transverse flagellum is reduced to a tooth-like projection.

The sulcus, however, is deep and prominent with a short flagellum. The most characteristic feature of the cell is a prominent tentacle extending posteriorly from the sulcus. Although the species lacks chloroplasts and is therefore holozoic, the cell has a pinkish hue and it is often a constituent of red tides as well as being one of the luminescent dinoflagellates. Cell division is by fission or multiple fission, producing numerous gymnodinoid swarmers which are thought to be gametes.

One family of holozoic, colourless dinoflagellates, the Warnowiaceae, is characterized by an ocellus, a complex lens, with an associated structure of pigmented granules (e.g. *Nematodinium, Erythropsidinium, Warnowid),* which may be involved in light perception.

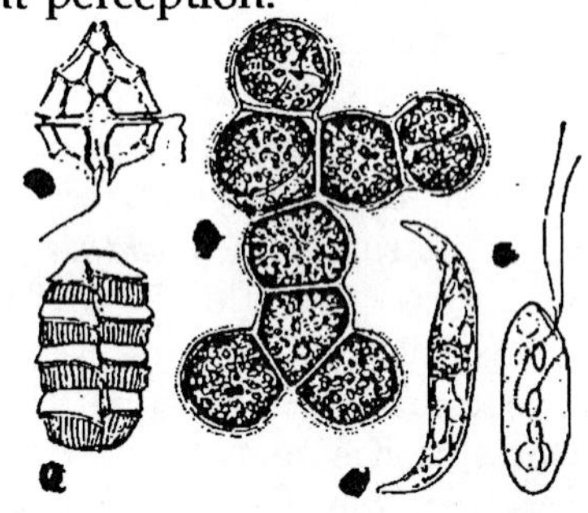

Fig. Polykrikos Schwartzii,

In the parasitic colourless Blastodiniaceae the majority are ecto- or endo-parasites of metazoa, though one species has been described as an ectoparasite on the diatom *Chaetoceras. Blastodinium* occurs as an elongate sac-like cyst in the gut of certain copepods. Within the cyst reproductive cells are formed initially from division of the sac cytoplasm.

After the first division only one of the resulting two daughter cells undergoes further division, so that eventually there remains one large daughter cell and numerous smaller ones. Upon rupture of the sac envelope the reproductive cells are liberated into the gut and the small ones divide once to give typical gymnodinoid zoospores which come to rest, secrete a new envelope, and thus start a new saccate individual.

The remaining large daughter cell from the first division may secrete a new envelope and in turn undergo a repeat division to give numerous small reproductive bodies. The ultimate fate of the large cell appears to be unknown.

PERIDINIALES

These are the armoured dinoflagellates and get their name from the heavy sculptured plates on the cell wall. The plates form the major distinguishing feature between the Peridiniales and the Gymnodinioideae. The number and arrangement of the plates is of great taxonomic importance.

The plates, which are usually arranged in three series on both the epitheca and hypotheca, are also present as a series of small ventral plates on the girdle and sulcus. The sculpturing on the plates is often very intricate, involving numerous tiny depressions (or areolae) with a minute hole, and often with teeth or spines. The girdle, which is usually very distinct, is commonly equatorial or nearly so, and is generally characterized by pronounced ridges on the edges.

In the Peridiniales the flagella emerge through a common slit rather than from two separate ones as in the Gymnodiniales. Also, in certain members the girdle spirals to the right rather than to the left. The majority of this order possesses chloroplasts, although holozoic nutrition is known to occur in some species. Trichocysts have also been recorded.

Reproduction is by oblique cell division of either motile or resting cells always through the point of flagellar insertion. The two parts of the old cell envelope may be retained by the daughter cells, or the protoplast may escape completely from the cell envelope, divide and then form a completely new theca, in which the old shell halves are left behind *(Peridinium)*. Sexual reproduction (e.g. *Ceratium)* and cyst formation (e.g. *Glenodiniurri)* have both been recorded for a few species.

Glenodinium, a genus of the predominantly freshwater Glenodiniaceae, is probably one of the simpler peridinioids. The cell shape is similar to *Gymnodinium,* but rudimentary armouring of small, delicate, indistinct plates is apparent. Sexual reproduction occurs in which a vegetative cell forms four gymnodinioid isogametes that fuse in pairs to form a resting zygote which eventually germinates to form four new flagellates.

Peridinium, family Peridiniaceae, is a common marine and freshwater genus with very highly developed sculpturing. The girdle is typically ridged and very prominent. A ventral concavity is often quite marked giving a kidney shaped appearance from top view.

In *Gonyaulax,* family Gonyaulaca-ceae, the two ends of the girdle are displaced; showing a spiralling tendency and the sulcus is usually quite distinct. As in *Peridinium* there is sometimes formation of antapical horns. In cell division, the separation line follows the sutures of the theca, and in some cases the daughter cells do not separate and chain-forming colonies may therefore occur.

G. polyedra is often the predominant species in red tides along California. *G. catenella* has been known to liberate a toxin in Californian mussels that can be fatal to humans, while *G. tamarensis* is a toxic species on the Atlantic coast. One of the most widely distributed and most specialized genera is the marine and freshwater *Ceratiwn.* The cells are highly asymmetrical and usually flattened dorsiventrally. The most characteristic feature is the elongated hollow horns.

As in most peridinioids (except *Glenodinium)* there is an apical pore, in this case at the tip of the apical horn. In the sulcal region is a large ventral area that is very lightly sculptured. Chains of individuals may be formed as a result of cell division. Sexual reproduction, involving amoeboid gametes has been reported and has recently been confirmed by Von Stosch. The female cells are similar to vegetative cells, but the male micro-swarmers are smaller.

DINOPHYSIALES

The genera of this order possess a cell wall comprised of two very large plates and numerous smaller ones. The order is included in the Dinophyceae, rather than Desmophyceae, because there is a girdle sulcus and theca composed of many definitely arranged plates. The epitheca is usually very small and lacks the apical pore.

The flagella orientation is similar to that of the Peridiniales. Reproduction is by longitudinal division, one half of the theca going to each daughter cell.

Many species appear to lack chloroplasts, though some species of *Dinophysis (phusa,* bladder) have them. In most members of the order the thecal margins around the girdle and suture lines of the valves are flared out to give a wing-like appearance. In nearly all instances the girdle is so far to the front that the wing-like projections give the appearance of a collar. In *Dinophysis,* which is flattened laterally, the upturned margins or wings on the girdle project anteriorly to give the collar a funnel shape.

DINOCAPSALES

This order represents the palmelloid condition in the Dinoflagellates and thus corresponds to the Tetrasporaceae, Chrysocapsaceae and Heterogloeales. There are but two monotypic genera in the order, *Gloeodinium montanum* and *Rufusiella insignis. Gloeodinium,* which is usually found in peat bogs, consists of two, four or eight subspherical cells enclosed in several gelatinous sheaths. The individual cells possess a large central nucleus and numerous small chloroplasts scattered throughout the cytoplasm. Reproduction is by cell division or by *Hemidinium*-like zoospores.

DINOTRICHALES

This is another small order with two genera, *Dinothrix* and *Dinoclonium (clonos,* confused), comprising those members that exhibit the filamentous habit. The plants consist of sparsely branched filaments each containing a few globose or cylindrical cells. The cells possess the typical large nucleus, together with numerous discoid chloroplasts. Reproduction is by gymnodinoid swarmers, formed singly or in pairs in the cells.

DINAMOEBIDIALES (RHIZODINIALES)

This comprises a single monotypic order with one marine

species, *Dinamoe-bidium varians.* As the name implies the vegetative phase is permanently amoeboid and the cells move by means of stumpy pseudopodia. The cells are holozoic, there being no chloroplasts, and there is a large characteristic nucleus. Cell division has never been recorded although encystment of the protoplast occurs within a gelatinous wall. Division of the cyst two or three times results in the production of short-lived gymnodinioid zoospores that soon revert to vegetative amoeboid shape.

PHYTODINIALES (DINOCOCCALES)

The members of this order are all immobile unicells (that is, they are in the coccoid condition) at least in the vegetative phase, and they have small chloroplasts. The majority is freshwater and the cells are either free-floating or attached.

There is no vegetative cell division, reproduction being entirely by two to four motile gymnodinioid zoospores, or two non-motile autospores. *Cysto-dinium (kustis,* bladder) is a free-living form with elongate, extended cells. *Pyrocystis (pyro,* fire) is a marine luminescent planktonic genus with large spherical cells, that reproduce by division of the protoplast into new individuals.

In *Pyrocystis lunula* (known as *Dissodinium lunula* or *Gymnodinium lunula)* the four to eight daughter cells elongate into crescent shaped cells which are liberated. These cells then produce eight gymnodinioid zoospores within the parent cell. Formation of the large spherical unicell from these zoospores has not been observed.

8
Chrysophyta Bacillariophyta

In the Pascherian classification the Xanthophyceae, Bacillariophyceae and Chrysophyceae were grouped into the one division Chrysophyta. Recent biochemical and electron-microscope studies have however prompted a re-evaluation of this concept. While there are some similarities there are also some distinct differences. The Xanthophyceae lack both chlorophyll *c* and the carotenoid fucoxanthin, and show some other pigment differences.

The Bacillariophyceae possess a very distinctive cell-wall structure in a silica skeleton. A second class, the Haptophyceae, has recently been split off from the Chrysophyceae because its members possess a distinctive third organ, the haptonema, and isokont flagella.

It is largely a matter of opinion whether the three or four classes are arranged into one, two or three divisions. The relative importance that is attached to the similarities and differences will be the deciding factor. We have retained three divisions with four classes although many workers still retain only the one division. It seems essential, however, that the Xanthophyceae, should be removed from the Chrysophyta,.irrespective of the position of the Bacillariophyceae.

The chloroplast pigments of the Chrysophyta and Bacillariophyta are similar. The products of photosynthesis are the glycosidic type reserves (chrysose, chrysolaminarin or leucosin) and oil. The flagellar arrangement (except in the Haptophyceae) is typically heterokont. Morphologically there is a range from the smallest flagellates to filamentous and colonial forms within the Chrysophyta. The Bacillariophyta are found only as single or colonial, non-flagellated unicells.

Most of the members are unicells but there are colonial

representatives and a very few are filamentous. Some representatives, such as the Silicoflagellates, have cell walls Impregnated with silica. A charactgjjsti^feature of the class is the endogenous cyst (statospore) with silicified walls of two equal or unequal partslmdthe opening closed by a plug. The motile cells possess one (pleuron-ematic, flimmergeissel) or two (flimmergeissel and peitschengeissel) flagella. The second flagellum is simple (peitschengeissel) and shorter. In some genera (e.g. *Mallomonas)* it is very greatly reduced, and is regarded as a photoreceptor.

Sexual reproduction has been infrequently receded, and is isogamous. Cyst formation commonly follows sexual fusiory; but cysts can be formed agamically. Most of the numerous speciesare freshwater planktonic, the great majority occurring intemperate or cold waters or else developing in summer when watef" tejnpejcatures rise. Many form a major component of the narmoplankton. There are some marine families.

Taxonomic treatment of the group has varied widely. Prior to the erection of the Haptophyceae five orders were generally recognized. These are the ChrySomonadales, Rhizochrysidales, Chrysocapsales, Chrysosphaerales and Chrysotrichales and they illustrate the evolution from flagellate to filamentous form. Bourelly suggested twelve orders. In 1962 Christen-sen removed those Chrysophyceae with isokont flagella (with or without an haptonema) to the Haptophyceae and the validity of this class is now generally recognized. The Chrysophyceae (as recognized by Christensen) is now subdivided into 5 orders.

Bourelly does not recognize the Haptophyceae, but divides the Chrysophyceae into 4 sub-classes: Aconto-chrysophycidae, Heterochrysophycidae, Isochrysophycidae, and Craspedo-phycidae. The Craspedophycidae are the colourless Choanoflagellates and are often assigned to a separate class, Craspedophyceae. They will not be considered here. The Isochrysophycidae, with two orders, correspond to the Haptophyceae. The other two subclasses correspond to the Chrysophyceae. The Acontochrysophycidae are those genera in which zoospores are lacking or in which the zoospores are amoeboid and non-flagellate. The Heterochrysophycidae comprises those genera with one or two heterokont flagella.

Other workers have adopted different interpretations. Bourelly recognized twelve orders, one of which has since been removed to the Haptophyceae. Fott and Round recognized five orders. The arrangement and number of orders can be expected to change as more work, especially with the electron microscope, is done. We have retained the five order classification (as in Fott and Round), but the student should be aware that other classifications also exist.

The five orders, Chrysomonadales, Rhizochrysidales, Chrysocapsales, Chrysosphaerales, Chrysotrichales or Phaeothamniales are analogous to the arrangement^in the'Chlofophyceae, Xa'nthophyceae, and Dinophyceae. In Bourelly's classification these five orders are combined into two, the Chromu-linales (uniflagellate motile cells) and Ochromonadales (biflagellate motile cells), both in the sub-class Heterochrysophycidae. These two orders in turn are divided into 3 suborders corresponding to the flagellate, coccoid or palmelloid and filamentous construction.

In three orders thereisjio rigidcell wall of a normalhype. The Chrysomonadales, which probably represent some of the more primitive types, are either unicells *(Ochromonas)* or else colonial sheathed forms *(Dinobryon, Synurd)* or peculiar organisms (Silicoflagellates) with an internal siliceous skeleton. The Rhizochrysidales include amoeboid forms with pseudopodia, that are naked *(Rhizochrysis)*, ensheathed *(Lagyniori)* or plasmodial *(Myxo-chrysis)*. The third order contains palmelloid *(Chrysocapsd)* or filamentous *(Hydrurus)* analogues of the Chlorophyceae.

The remaining two orders comprise species with a rigid cell wall, e.g. the coccoid Chrysosphaerales and the filamentous Phaeothamniales. This last order is of particular interest in relation to the origin of the Phaeophyta.

The pigmented Chrysophyceae are holqphytic but there are colourless Chrysomonads, recognized by the typical plugged, and these are heterotropijic, the organic carbon being obtained by diffusion (osmotrophic) or ingestion (holozoic).

CHRYSOMONADALES

This order represents the simplest types, simple flagellates with two, rarely one or four, and flagella. The order may be sub-divided into two sub-orders, the Chrysomonadineae and Silicoflagellineae or Silicoflagellates. More recent proposals have suggested three orders: Chromulinales (uniflagellate), Ochromonadales (heterokont biflagellate) and Dictyochales (Silicoflagellates). In this latter arrangement the Ochromonadaceae, Synuraceae and Dino-bryaceae are referred to the Ochromonadales, while the Chromulinaceae is assigned to the Chromulinales.

Multiplication commonly takes place by cell division and characteristic cysts are formed during unfavourable conditions. Sexual reproduction has been reported in a few instances.

OCHROMONADACEAE: OCHROMONAS

The species are mostly planktonic or attached, freshwater organisms, each individual containing one or two chloroplasts with pyrenoids. There is a contractile vacuole and a red eyespot. Flagellar structure is typically heterokontic, with fibres passing from the flagellum base to the side of the nucleus. Reproduction takes place by simple fission but siliceous cysts are formed at the end of the vegetative period.

Palmelloid stages have been observed in some species. Nutrition of the species varies from phototrophic to phagotrophic to saprotrophic. In some genera cells like those of *Ochromonas* are united into colonies joined by stalks, e.g. *Chrysodendron*, or enclosed in a jelly, e.g.

Uroglena. *Amhophysa* is a widespread colourless representative. *Didymochrysis* is quadriflagellate, and uninucleate and looks like a monad that has failed to divide.

CHROMULINACEAE: CHROMULINA (CHROMULINA, YELLOW).

Whereas *Ochromonas* is a typical representative of the biflagellate group (Ochromonadales), *Chromulina* may be considered typical of the uniflagellate group (Chromulinales). *Chromulina* is a large genus of both marine and freshwater species. The cells typically contain one or two pyrenoid-bearing chloroplasts, an eyespot and contractile vacuole. The cell membrane is non-rigid to allow metabolic movements. The single flagellum is pleurone-matic. Reproduction is by cell division. Smooth-walled cysts also occur.

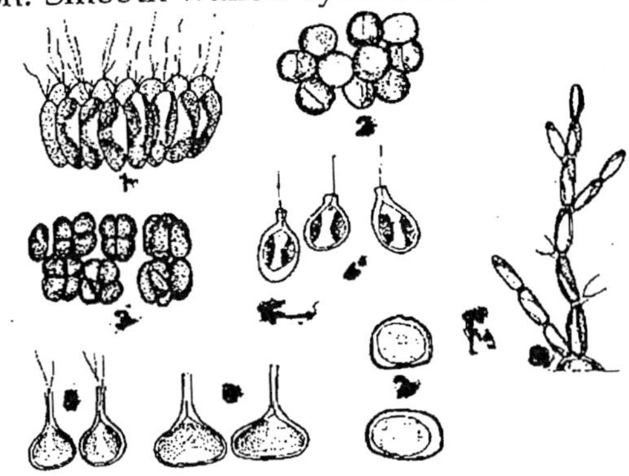

Fig. Chrysophyceae.

DINOBRYACEAE: DINOBRYON

The members of this family all exist in a cellulosic sheath (lorica) formed from the protoplasm, which is either fine and colourless or thick and impregnated with iron. In some genera the cells are single but in *Dinobryon* bushy colonies are formed by a series of empty sheaths. The cells possess two chloroplasts and contractile vacuoles, an eyespot and heterokont flagella. The cells are attached to the inside of the lorica by a contractile stalk. When longitudinal fission occurs two swarmers emerge at the orifice and two new sheaths are formed.

Typical cysts with plugs are formed and as some of these are binucleate they presumably arise as a result of cell fusion. Some species can be abundant in marine and fresh water, but in the latter case they are mainly found in oligosaprobic waters. *Lepochromulina* is attached to the substrate by the broad base of its sheath.

SYNURACEAE: SYNURA

This is a floating colonial genus in which the sheaths are united by a pectinaceous material. The individual monads each have two flagella, which are slightly unequal, but heterokont and heterodynamic. The longer flagellum points to the front, the shorter to the rear.

There are two chloroplasts in each cell and a number of contractile vacuoles. The periplast is overlaid by a layer of finely sculptured scales which are both ellipsoidal and siliceous. The monad cell may divide within the colony or without to form a new colony.

In *Mallomonas,* some species of which are colourless, the cells are free and solitary with siliceous scales that often possess a distinct spine. The second flagellum is greatly reduced, often being barely discernible and is regarded as a photoreceptor. Besides cell division and asexual zoospore formation, isogamous sexual reproduction has been described.

SILICOFLAGELLINEAE: DICTYOCHA

The silicoflagellates constitute a sub-order which is sometimes assigned ordinal status as the Dictyochales. They are phototropic organisms found in marine plankton and possess an internal siliceous skeleton. This skeleton preserves well and hence many genera are known in the fossil condition from the Tertiary onwards. The body is naked and extrudes pseudopodia. There is a single flagellum and the chloroplasts are either in the main body or else in the outer plasma tic layer.

RHIZOCHRYSIDACEAE: RHIZOCHRYSIS

These amoeboid organisms with long pseudopodia possess one or two typical chloroplasts but lack flagella cr an eyespot. During division a daughter cell may fail to obtain chloroplasts and thus produce a colourless individual. *Chrysamoeba* is a very similar genus but the cells differ in possessing a short flagellum so that the genus is sometimes grouped with the Chrysomonadales.

The Lagynionaceae contain all those sheath-living members that represent amoeboid analogues of the Dinobryaceae. The peculiar species *Myxochrysis paradoxa* (Myxochrysidaceae) produces a multinucleate plasmoidal vegetative form with chloroplasts, contractile vacuoles and chrysose. Nutrition is achieved by digesting small unicellular algae. Like *Chrysamoeba* this genus is sometimes referred to the Chry somonadales.

During reproduction the plasmodium forms a number of cysts which produce chrysomonad-like flagellates. These lose their single flagellum, become amoeboid and fuse with each other to form a new plasmodium.

CHRYSOCAPSALES

In this order, representing the colonial habit, the cells are contained within a gelatinous matrix. In the vegetative phase the cells lack a flagellum but they do possess chloroplasts and chrysose and many also have contractile vacuoles. There are two families: the Chrysocapsaceae, in which the cells are in irregular or spherical gelatinous masses, e.g. *Chrysocapsa,* and the Hydruraceae in which the gelatinous material is a simple or branched, erect thallus with an apical growing point. Bourelly does not recognize the order and assigns both families to the Chromulinales.

In *Hydrurus foetidus* the vegetative cells, each oriented in the same direction, contain one hemispherical chloroplast with a single pyrenoid, four contractile vacuoles and a large chrysose ball. Vegetative growth of the plant, which occurs in quickly-flowing fresh water, takes place by means of the apical or branch terminal cells. Reproduction is secured by means of tetrahedral zoospores produced singly per cell; each has two flagella, one long, the other short and thick. The long flagellum is locomotory and the short is used to give direction. Siliceous cysts are also produced.

PHAEOTHAMNIALES

This order contains, because of their filamentous organisation, the most highly evolved members. They can be regarded as derived from the previous coccoid order, although Bourelly considered that the coccoid form was reduced from the filamentous. The plants are simple or branched erect *(Phaeothamniori)* or prostrate *(Phaeodermatium)* threads. The prostrate forms may develop into a pseudo-parenchymatous layer. Reproductive swarmers are of the *Chromulina (Phaeodermatium)* or *Ochromonas (Phaeothamniori)* type. The species are not so common in nature as other Chrysophyceae but they arc both marine and freshwater. There are four families each with but few genera.

In the Phaeothamniaceae, as represented by *Phaeothamnion (Phaeo,* brown; *thamnion,* thread), there is a branched filament arising from a hemispherical basal cell which represents the remains of a typical cyst. When reproducing, four to eight *Ochromonas-Vike* swarmers are produced per cell. Creeping, branched forms, that later form a pseudo-parenchymatous layer, and which reproduce by *Chromulina-like* zoospores are placed in the Phaeodermatiaceae. In *Phaeodermatium* the small discs are found on stones in cold, quickly flowing waters. The remaining two families are not well known.

HAPTOPHYCEAE

This is a relatively new grouping of certain unicellular flagellates which are predominantly marine. Recent electron microscope studies showed that certain flagellates originally assigned to the Chryso-phyceae, often possessed a third whip-like organ in addition to the two flagella. This special organ is called a haptonema, and the terminal portion or the whole organ can contract into a tight spiral and attach the cell to a substrate. The two flagella, in contrast to the Chrysophyceae, are equal, that is, they are isokont.

Electron microscopy has also shown that the organisms are covered with minute scales very like those recorded for the Prasinophyceae. These flagellates were originally grouped in the Chrysophyceae to which they are closely allied. Each cell normally contains two brown parietal chloroplasts, with Chrysophycean pigments and oil droplets associated with the typical chrysophycean reserve product, leucosin. Some workers (e.g. Bourelly) retain this group within the Chrysophyceae, as a sub-class Isochrysophycidae.

There are two orders, the Isochrysidales and the Prymnesiales. In the former, the motile phase lacks the haptonema, whereas motile cells of the Prymnesiales possess the haptonema. This latter order includes a group of planktonic organisms known as the Coccolithophorids which possess mineralized scales of two types.

The holococcoliths are composed of submicroscopic crystals, and such scales or crystalloliths are found on *Crystallolithus hyalinus* which is the motile phase of *Coccolithus pelagicus*. The hetero-coccoliths are larger with more obvious structures built up into plates, ribs or girdles.

Many of the known representatives of the division are not well understood. Until further work has been carried out on certain genera their position in this division or in the Chrysophyceae must remain provisional.

ISOCHRYSIDACEAE: ISOCHRYSIS

The individual cells which are naked and capable of metabolic movements are pyriform in shape with two typical chloroplasts placed laterally on each side and with a red stigma at the anterior end. Reproduction takes place by the formation of an endogenous cyst which, on germination gives rise to zoospores or a palmelloid phase. There appears to be some form of primitive sexuality in which two non-motile cells fuse, after which one nucleus disappears and divisions take place giving four daughter cells. Each of these cells produces a cyst and from this a new motile cell eventually emerges. The alga will grow rapidly in saline concentrations ranging from 1-5-4 per cent.

PRYMNESIACEAE: CHRYSOCHROMULINA

The original representative of this genus was freshwater but now a number of marine species have been described by Parke and her co-workers. In all cases the two flagella and the haptonema are very long, the latter usually being longer and thinner than the flagella. The haptonema often has a coiled appearance. The tip of the haptonema is adhesive and by this means the organism can become temporarily anchored. In all cases the haptonema consists of three concentric membranes enclosing a ring of six *(C. strobilus)*, seven (most frequent), or rarely eight (C. *kappa)* fibres or tubes.

This variation in number of fibres is a distinct contrast to the fixed number that is found in flagella of other groups. All the known species possess a covering of scales, which are apparently of an organic nature, and if it were not for their small size they could be used for taxonomic purposes. In C. *ericina,* the scales have spines while in *C. strobilus* two distinct types are arranged in separate layers. In C. *pringsheimii,* there are four different kinds of scales, two plate-like and two with spines. In the motile phase the chloroplasts are saucer shaped whereas in the non-motile phase they become stellate.

The cells possess peripheral muciferous organelles which are capable of expelling their contents. In the motile phase reproduction takes place by means of ordinary fission, the two daughter cells usually being of equal size. In the non-motile phase the cells become amoeboid and may fuse since four plastids can be seen in some larger organisms. Division then occurs and four non-motile daughter cells are produced from each of which a new motile cell escapes. Ingestion of bacteria and other small algal cells by phagotrophy takes place at the non-flagellar end of motile cells. Empty walls of ingested cells are extruded through temporary colourless tubes.

In *Prymnesium parvum,* the haptonema is much shorter and does not coil. As in *Chrysochromulina* the cell membrane is covered with numerous unmineralized scales. Muciferous bodies are located just inside the membrane. Reproduction is by cell division. Production of cysts has been recorded in natural populations. This flagellate synthesises a variety of ichthyotoxins which cause problems in Israeli fish ponds.

PHAEOCYSTACEAE: PHAEOCYSTIS

The species are planktonic or epiphytic. The individual ovoid or spherical cells are distributed irregularly around the periphery of a gelatinous spherical or irregular colony. Each contains one to four yellow plastids arranged parietally. Reproduction is by means of zoospores with two equal or unequal flagella. In P. *globosa* there is a third short organ (possibly a haptonema). The most important species is *P. poucheti,* the 'baccy juice' of fishermen. This alga, a common food of antarctic Crustacea, and thus penguins, synthesises significant amounts of acrylic acid which results in a greatly reduced intestinal flora in the penguins.

Herrings are repelled by this organism when it is present in mass, and a vernal maximum off the coast of the Netherlands turns the northward herring migration west towards Britain, and brings about the spring fishery off the coast of East Anglia. The occurrence of an abnormal autumn maximum out of its usual station may completely change the grounds of the autumn fishery during the southward migration. The organism also forms a bloom in the Irish Sea, especially in Liverpool Bay, where it does not appear to have any effect upon other planktonic forms.

It has been suggested that the inshore occurrence of such blooms means that some factor associated with terrestrial drainage is mainly responsible. Further work on this organism is clearly necessary because of its economic importance and also because there must be some dormant phase that has not yet been described.

COCCOLITHACEAE: COCCOLITHUS

These small organisms form part of the nannoplankton and have been known for a long time. *Coccolithus huxleyi* is the commonest coccolithophorid. The characteristic feature is the presence of the holo- and heterococcolith scales. The holococcoliths are composed of a delicate scale comprising rhombohedral or hexagonal calcite crystals in a regular array. Such coccoliths are found on the motile stage *(Crystallolithus hyalinus)* of the non-motile *Coccolithus pelagicus.* The more abundant heterococcoliths are found in *Coccolithus huxleyi and* the non-motile stage *of Coccolithus pelagicus.*

These placoliths, also built of rhombohedral crystals, are composed of two flattened discs joined by a short tube. In *Coccolithus pelagicus* (non-motile) the placoliths appear to be attached on the inner disc to a non-mineralized scale. As with the scales of *Chrysochromulina* the coccoliths are formed within the golgi apparatus and extruded to the surface. Meanwhile the old external coccoliths are shed. Like other Prymnesiales, the flagellate cells possess two smooth flagella and a haptonema. The chloroplasts possess the typical Chrysophyceae-Haptophyceae pigments. Chrysolaminarin is the storage reserve. The siliceous cysts typical of the Chrysophyceae are unknown and reproduction is by cell division.

Coccolithus pelagicus demonstrates an unusual life history. A small unicellular motile state, which normally reproduces by division can in older cultures give rise to a large unicellular non-motile stage *(Coccolithus pelagicus).* These non-motile cells then divide twice to give rise again to the motile 'Crystallolithus' stage. In some genera two kinds of coccoliths can be found on the same cell, e.g. *Michaëlsarsia.* In the Syracosphaeraceae (e.g. *Cricosphaera carterae)* the coccoliths are of the heterococcolith type (Cricoliths). These are very similar to those of *Coccolithus pelagicus* and as in this organism, unmineralized scales have also been recorded.

In *Cricosphaera,* however, the motile stage divides twice to produce a tetrad, which further divides to give rise to multicellular filamentous stages which lack coccoliths and resemble such Chrysophycean algae as *Chryso-sphaera* and *Thallochrysis.* It has been demonstrated that the motile Coccolith stage of *Cricosphaera carterae* is diploid and the benthic filamentous stage is also diploid; that is, a heteromorphic alternation of generations.

BACILLARIOPHYCEAE

These unicellular algae, the diatoms, are abundant as isolated or colonial forms in marine or freshwater plankton, and also as epiphytes on other algae and higher plants. They form a large proportion of the bottom flora of lakes and ponds and occur widely on salt marshes, although certain diatoms are said to be very sensitive to the degree of salinity in the medium. Diatoms are also found in the soil where they can form quite a substantial proportion of the soil flora.

In the rain forests of the tropics they can be associated with Cyanophyceae on the leaves of trees. Most species exist as single cells but there are some colonial species in which the cells are attached to each other by mucilage or else are enclosed in a common mucilage envelope.

The Bacillariophyceae were originally classified with the Chrysophyceae in the Chrysophyta, as there are many important similarities between the two groups. The chloroplasts are olive green to brown in colour and possess chlorophylls *a* and *c* with fucoxanthin and acetylenic carotenoids. The products of photosynthesis are oil and chrysolaminarin.

The gametes possess a pleuronematic flagellum and silicification is present on the cell wall. The most characteristic feature of the class is the highly ornamented and sculptured silica wall, overlaying an inner pectin layer and organic matrix.

The diatoms are generally classified into two orders, the Pennales and Centrales. The Pennales are characterized by a bilateral symmetry, their gliding motion and sexual reproduction (when present) by an amoeboid isogamy. The

Centrales are radially symmetrical, lack the gliding motion and demonstrate anisogamy or oogamy with motile male gametes. Floating and predominantly marine forms are found in the Centrales as against mainly mud, soil and epiphytic forms in the Pennales. The further classification is based almost solely on the structure of the siliceous walls, which are highly characteristic and magnificently sculptured.

Each shell or frustule is composed of two halves of varying shape, the older and larger epitheca, fitting closely over the younger and smaller hypotheca. Each half consists of a valve with a connecting band. The connecting band of the epitheca overlaps that of the hypotheca and the overlap region is referred to as the girdle. Consequently a dorsal or ventral view is referred to as a valve view; a side-on view is called a girdle view. The whole arrangement can be pictured as resembling a pill box or petri-plate.

The shells in nearly all cases contain many tiny pores; these perforations (punctae) which are of a very complex structure form from 10-30% of the valve area. These pores can be more or less occluded by outgrowths or by thin perforated siliceous membranes. Often the punctae are very closely spaced in linear array (striae) giving the cell a striated appearance. The striae are either radial from a central point (Centrales) or in longitudinal rows. In jhe Pennales the two rows of striae on each valve may be separated by a cleft or groove (raphe) or a clear area (pseudoraphe). In transverse section the raphe appears as a V-shaped fissure, with the two arms of the fissure opening to the inside and outside of the wall.

In the Pennales the raphe can be one of two kinds. In the one it consists of two medium longitudinal slits following a straight or sinuous course, while in the other (canal raphe) it is a canal lodged in a crest or keel. At each pole of the cell, adjacent to the end of the raphe there is a nodule or thickening of the valve wall (polar nodule). A similar thickening occurs in the centre of the cell (central nodule) and this divides the raphe in half.

The modifications of this general pattern of the silica-wall structure are multitudinous. Within the Centrales various morphological forms are represented. Cells in valve view may be circular or triangular *(Triceratium)*. In girdle view the cells may be appressed *(Coscinodiscus, Cyclotelld)* or cylindrical *(Melosira, Biddulphia).*

In some cases (e.g. *Chaetoceros)* long horn-like projections extend from the four corners in girdle view. Centric diatoms may be solitary *(Coscinodiscus)* or form chains *(Melosira, Chaetoceras)* and in occasional instances form colonies. *Coenobiodiscus* produces a uniseriate colony *ofThallasiosira-like* cells all interconnected by a filamentous matrix. The Pennales are often subdivided according to the presence or absence of raphe and pseudoraphè:

The Araphidineae (e.g. *Rhabdonema, Thalassiothrix)* have a pseudoraphe, while the Raphidioidineae (e.g. *Eunotid)* have a rudimentary raphe. If there is a raphe on one valve and a pseudoraphe on the other the genera are assigned to the Monoraphidineae (e.g. *Cocconeis).* The Biraphidineae (e.g. *Cylindro-theca, Phaeodactylum, and Gomphonemd)* houses those genera with a raphe on both valves. Cell form may range from oval *(Cocconeis),* boat-shaped *(Phaeodac-tylum)* and long and narrowly fusiform *(Nitzschid). Phaeodactylum* is noteworthy in being very weakly silicified.

Colonial forms as filaments or regular aggregates the pennate diatoms that have a raphe are capable of movement but a great deal more work is necessary before the mechanism is fully understood. There may indeed be more than one type of mechanism, especially as some diatoms can move when either the girdle or the valve is in contact with the substrate. A recent important study by Drum and Hopkins has shown that linear streaming always in the opposite direction to that of locomotion is necessary in one or more raphe systems and that movement cannot occur without adhesion at a point of contact between the raphe system and the substratum. A secretion is produced during locomotion and this forms a mucous trail on the substrate.

The nature of the mucoidal material is not known. The source of the secreted material is crystalloid bodies found only in the cytoplasm of motile diatoms. These motile diatoms have a fibrillar contractile system that is capable of continuous, rhythmic contraction and which produces the necessary pressure for forced secretion and streaming. This fibrillar system is activated by a pjiotophobic, phototactic, mechanical or nutritional stimulus.

Cell division normally occurs at night and when the nucleus and protoplast have divided new valves are formed inside and then the parent connecting bands separate. The two frustules of each parent cell act as the epitheca for the two new cells. The size of the daughter cells is thus determined by the size of the frustules of the parent. Consequently there is a dimunition in size of successive generations. The average species size is restored by auxospore formation. There are some species that do not decrease in size and it is thought that in such cases the plasticity of the cell wall and the construction of the girdle in the form of an open hoop allow cells to resume their original size on division.

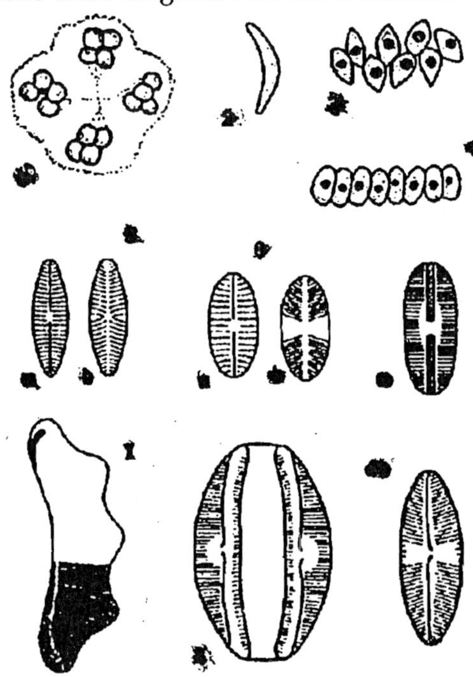

Fig. Bacillariophyceae

The vegetative cells of all diatoms are diploid and meiosis occurs at gamete formation. Apart from *Rhabdonema adriaticum,* all species studied are monoecious.

The centric diatoms are oogamous, the young oogonia

being recognized by increase in size of nucleus and extension of the cell. Normally only one ovum matures while the other three nuclei formed during meiosis disintegrate. In *Isthmia* the nucleus divides in two and one nucleus aborts. The remaining nucleus again divides, with one of the daughter nuclei degenerating.

In *Biddulphia mobilensis* two ova are produced per oogonium. Sperma-togonia can be formed in a variety of ways:

1. A vegetative cell may become a spermatogonium directly (Cyclotella tenuistriata);
2. A vegetative cell may divide to give several spermatogonia (Melosira varians);
3. Spermatogonia may remain enclosed within the mother cell or they may complete their development after liberation. Each spermatogonium, after meiosis, produces four spermatozoids, each with a single flimmergeissel flagellum in which'the two central strands may be missing. Fertilization occurs in the female cell (auxospore mother cell).

In those centric diatoms where no sex cells have yet been seen it is assumed that sexual reproduction is by means of intracellular autogamy. In this case (e.g. *Cyclotella),* the nucleus of one cell divides to give four haploid nuclei, two of these degenerate, while the remaining two fuse within a mass of protoplasm to give the 'zygote'.

The pennate diatoms (e.g. *Cocconeis, Surirelld)* are isogamous and the gametes amoeboid. Usually two or three of the four meiotic nuclei degenerate. The cells come together in pairs and connect through a conjugation tube. The female gamete usually remains in the parent cell. The frustules of the male cell separate and the male gamete migrate to the female cell where fertilization occurs. In *Gomphonema* autogamy has been reported. In this case two nuclei from meiosis degenerate. The protoplasm divides to give two 'gametes' each with one undegenerated and one degenerated nucleus. These fuse within the cell.

In both diatom groups the zygote starts to grow at once into an auxospore, which eventually becomes about three times the width or length of the mother cell. Two mitotic divisions then occur, at each of which one daughter nucleus aborts. A new shell is formed after each division.

In *Rhabdonema adriaticum*, the sole oogamous pennate diatom, only one ovum is produced per oogonium. The male mother cells produce spermatagonia that are dispersed passively in the water and become attached by mucilage pads to the oogonia. They then give rise to naked, non-flagellated microgametes from which only the nucleus enters the ovum. It is thought that this species may represent the kind of transitional type that formerly linked the centric and pennate diatoms.

Phylogenetically the centric diatoms are regarded as more primitive and they also have an older geological history, occurring first in the Jurassic. Among the pennate diatoms the Araphidineae (those with only one pseudo-raphe) provide some transitional stages. They presumably evolved into the Monoraphidineae and then the Biraphidineae.

It is quite clear from the single pleuronematic flagellar structures, and from wall silicification and pigment composition, that there is a close relationship between this division and the Chrysophyta. There is rather less affinity with the Xanthophyta.

PHAEOPHYCEAE

The algae composing this class range from minute discs to thalli 100 metres or more in length and are characterized by the presence of the carote-noid, fucoxanthin, which masks the chlorophyll. The class can be divided into a number of orders and families which can be treated independently, or the families may be placed into three groups as proposed by Kylin.

These groups are based upon the type of alternation of generations, though the classification involves difficulties so far as the families Cutleriaceae and Tilopteridaceae are concerned and it is therefore best not to use a formal classification. The three groups are:

Isogeneratae

Plants with two morphologically similar but cytologically different generations in the life cycle (e.g. Ectocarpaceae, Sphacelariaceae, Dictyptaceae)..

Heterogeneratae

Plants with two morphologically and cytologically dissimilar generations in the life cycle: (i) Haplostichineae: Plants with branched threads, which are often interwoven, and usually with trichothallic growth (e.g. Chordariaceae, Mesogloiaceae, Elachistaceae, Spermatochnaceae, Sporochnaceae, Des-marestiaceae). (ii) Polystichineae: Plants built up by intercalary growth into a parenchymatous thallus (e.g. Punctariaceae, Dictyosiphonaceae, Laminariales).

CYCLOSPOREAE

Plants possessing a diploid generation only (e.g. Durvilleales, Ascoseirales, Fucales).

In any phylogenetic consideration there is really very little difference between the two classifications. Most of the recent significant proposals have involved revision at the class and division levels, and there have been few significant changes in the lower taxa.

The Phaeophyceae are extremely widespread and are confined almost entirely to salt water, being most luxuriant in the colder waters, though the genera *Heribaudiella, Pleurocladia* and *Bodanella,* and *Sphacelaria fluviatilis* occur in fresh water. All of these except the *Sphacelaria* belong to the Ecto-carpales; many of the records come only from China.

Some species exhibit morphological variations and it has been shown that these may depend on both the season of the year and the nature of the locality. Although the unicellular and siphonous habits are unknown, extremely varied morphological forms do occur, and those range through simple filament *(Acinetospora),* heterotrichy *(Myrionema, Scytosiphori),* interwoven filaments or cable type *(Myriogloia),* hollow parenchymatous sphere *(Lea-thesid),* cortication *(Sphacelaria),* and simple *(Punctarid)* or complex parenchymy (Laminariales, Fucales). The large size of many Phaeophyceae and their location in the surf zone require additional mechanical support. This can be seen in *(a)* increased wall thickness *(Stypocaulori), (b)* twisting and rolling together of threads, (r) occurrence of anchorage devices ('root' branches) or *haptera.*

Growth in the primitive brown algae is by an intercalary

meristem at the base of a hair or filament growth. In some of the advanced orders trichothallic growth has been replaced by an apical cell or group of apical or marginal cells. Hyaline hairs occur in many species. Cells vary greatly in size but they always have distinct walls, which are usually composed of cellulose on the inside and pectin outside, and although they are usually uninucleate, they occasionally become multinucleate.

When the fucoxanthin-protein complex of the plastids has been denatured the thallus becomes green. Usually each cell contains several chloroplasts, which are commonly parietal and discoid though there is some diversity of form. Although pyrenoids have been described for some species they are not wholly comparable with those in other algae as they appear as peripheral bulges on the chloroplast. Chlorophyll b is replaced by chlorophyll c, but unlike the other Chromophyta these algae have no acetylenic carotenoids.

The products of assimilation are sugar alcohols, fats and the complex polysaccharides, laminarin and fucoidin. A characteristic feature of the Phaeophyceae is the presence of colourless, highly refractive vesicles known as fucosan vesicles or physodes. These contain a phenolic material known as fucosan, and they are particularly abundant in tissues where active metabolism or division is taking place.

Vegetative reproduction may take place by splitting of the thallus or else by the development of special propagules. In some orders perennation is achieved by means of a perennial, prostrate attachment organ. Asexual reproduction is commonly secured by means of biflagellate, typically heterokont zoospores which are normally produced in specialized sporangia. In the Dictyotales tetraspores replace zoospores. In Zonaria, and at times in Pocockiella, eight spores are produced instead of the usual four. In the Tilopteridales asexual reproduction is by means of uni- to quadrinucleate monospores.

Monospores are best regarded as forerunners of tetraspores, especially when considered in relation to the vegetative characters. Sexual reproduction ranges from isogamy (with both pear-shaped gametes characteristically bearing two flagella inserted laterally), to anisogamy and finally to oogamy.

The ova are not normally retained on the parent plant and fertilization therefore takes place in the water, though the ovum may remain attached by a long mucilaginous stalk. The change from isogamy to anisogamy is accompanied by a corresponding differentiation of the gametangia.

Electron-microscope studies of the antherozoids of several species have revealed the fact that the flagellum possesses the typical 9+2 arrangement. The longer flagellum is of the flimmergeissel type (pleuronematic or panto-nematic). The shorter one trails and is a peitschengeissel type, although in some Fucales, the reverse situation occurs. In some Fucalean antherozoids (Fuca-ceae) there is a peculiar transparent beak-like proboscis which is part of the flagellar apparatus. Another feature in the antherozoids of *Himanthalia* and *Xiphophora* is a single spine on the front flagellum near the distal end. There is morphological evidence that the uniflagellate *Dictyota* antherozoids have been derived from biflagellate ancestors.

Plurilocular organs can be borne on both diploid and haploid plants, whereas unilocular sporangia are normally confined to diploid plants. Plurilocular sporangia found on diploid plants represent an accessory means of reproduction, while those found on the haploid generation are gametangia. In isogamous forms gametes are produced from plurilocular structures, so that in anisogamous forms the antheridia and female gametangia must be regarded as modified plurilocular structures.

In general the presence of a unilocular or tetrasporic sporangium indicates a diploid thallus, and as meiosis occurs in the formation of the contents the swarmers are invariably haploid. The chief evolutionary features to be observed within the Phaeophyceae, at least as far as reproduction is concerned, are:

1. The elimination of accessory reproduction (the plurilocular sporangia) in the sporophyte.
2. A tendency to soral aggregation of sporangia on the sporophyte, e.g. Sphacelariales, Dictyotales, Laminariales.
3. A tendency towards anisogamy and finally oogamy.

4. A tendency towards an increase in the basic haploid number of eight chromosomes.

Most species show an alternation of generations, but this is by no means regular, and the phenomenon in some orders at least is therefore better termed a life history.

In some orders (e.g. Laminariales) a very marked heteromorphic alternation of generations exists. Within the Heterogeneratae, where the game-tophyte is much reduced, the regular alternatioil may be masked by such phenomena as parthenogenesis, e.g. *Laminaria, Alaria*. Those members of the Heterogeneratae (excluding the Laminariales) which exhibit this type of alternation have a fully developed diploid or delophycee form which is common in summer, and a much reduced haploid or diploid adelophycee stage which usually appears during the winter months.

ECTOCARPALES

The members of this order, which are the least specialized of the Phaeo-phyceae, are essentially branched heterotrichous filaments with prostrate and erect portions of the thallus; one or other portion may subsequently become lost as in *Phaeostroma*. A number represent forms that have become reduced as a result of either an epiphytic or parasitic habit. Despite the views of Russell it is regarded here as more desirable to retain this order in a restricted sense and to exclude plants that can be placed in the Chordariales and Dictyosiphonales.

ECTOCARPACEAE: ECTOCARPUS

The plants are composed of sparsely or profusely branched uniseriate filaments. The aerial portion arises from a rhizoidal base, which in some epiphytic species occasionally penetrates the host. *E.fasciculatus* grows on the fins of certain fish in Sweden, but the nature of this relationship is not clear. The branches of certain genera terminate in a colourless hair; in young plants of *E. siliculosus* pseudo-hairs^may occur, but later, with increasing age, they disappear through truncation. The erect filaments of some species have an intercalary growing region, while in other species growth is diffuse; the rhizoids increase in length by means of apical growth.

Each cell, which contains one nucleus together with brown, disc or band-shaped chloroplasts, possesses a wall that is composed of three pectic-cellulose layers. Generally two kinds of reproductive structures are present, the plurilocular gametangia and sporangia and unilocular sporangia. The unilocular structures always occur on diploid plants and they give rise, after meiosis, to numerous haploid zooids which may either function as gametes or else develop without undergoing a fusion.

The sporangia are sessile or stalked and vary in shape from globose to ellipsoid, the mature ones dehiscing through the swelling in the centre layer in the wall. The plurilocular structures, which are either sessile or stalked, range from ovate to siliquose in shape and are present on haploid or diploid thalli. In *E. siliculosus* they represent modified lateral branches. The plurilocular organs are divided into a number of small cells, each one of which gives rise to a swarmer.

When the swarmer is ripe, dehiscence takes place by means of a pore and the contents either germinate directly or else behave as isogametes, depending on the genetic status of the plant. Recently Miiller has shown that what are probably female gametes exude a volatile odiferous substance to attract gametes of the opposite sign.

In *Giffordia secunda (E. secundus)* there is well-marked anisogamy, the smaller gametangia being antheridia and the larger female gametangia. The contents of the latter are sometimes capable of parthenogenetic development. In *E. padinae* the unilocular sporangia are absent and there are three kinds of plurilocular organs. One type, which has very small loculi, represents the antheridia, while there are also medium-sized or [4]meiosporangia and larger organs. One of these must represent the female reproductive organs.

The life histories of the species are full of interest, especially in view of what has been discovered for *E. siliculosus.* Knight found that the plants in the Isle of Man occurred in early spring and late autumn and were all diploid. They bore unilocular and plurilocular structures, the former producing gametes while the latter gave rise to zoospores. In the Bay of Naples, on the other hand, the plants were all haploid and only bore plurilocular organs. The zooids from these behaved as gametes. Berthold recorded a microscopic form which has since been regarded as diploid because unilocular sporangia were found on it.

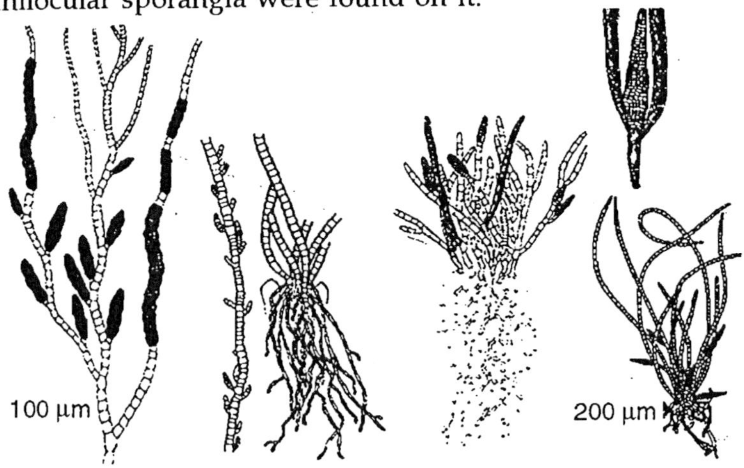

Fig: Ectocarpus

Subsequently Schussnig and Kothbauer and Miiller revealed the existence of unilocular structures. It would seem, therefore, that the somewhat complex schema of these later workers is the more correct, at any rate so far as the Neapolitan form is concerned.

It has been suggested that the differences between plants from the two localities are due to differences in the tides, light conditions or temperature, with perhaps more emphasis on the last. Muller has since shown that in one form of *E. siliculosus* unilocular structures are formed at 13°C and plurilocular at 19° C and that there is a lunar periodicity in the liberation of the swarmers.

The later results indicate that the earlier work of Knight in the Irish Sea needs to be repeated. Yet another study of this species has been carried out in America by Papenfuss. In America the diploid plants were found growing epiphytically on *Chorda* or *Spartina* and these either bore pluri- or unilocular structures independently, or else both could be found on the same thallus.

The unilocular structures occurred only in summer while the plurilocular were present throughout the whole year.

Although the zooids from both types of structure acted as zoospores and germinated directly, nevertheless meiosis always took place in the unilocular structures, the zooids of which developed into the sexual plants that were found growing as obligate epiphytes on *Chordaria,* in some cases the earest asexual plants being twenty miles distant. The plants growing on *Chordaria* were dioecious and only bore plurilocular gametangia.

This extremely large genus is now subdivided, and a number of new genera have been established, leaving two main species complexes, *E. siliculosi* and *E.fasciculati* within the genus. The view of Russell that the *E. siliculosus* complex represents a clone is in agreement with the variations found in the life histories.

The allied genus *Pylaiella* is very similar to *Ectocarpus* morphologically. Two species, *P. littoralis* and *P. (Bachelotid) fulvescens* are widespread; the former occurring often as an epiphyte whiles the latter is generally saxicolous. The species are readily distinguished from *Ectocarpus* by the position of the sporangia because these bodies are nearly always intercalary, very rarely terminal, and when this latter is the case it is frequently due to the loss of the terminal vegetative portion. A study of British populations has shown that there are two entities present that may differ genetically and physiologically.

TILOPTERIDACEAE: HAPLOSPORA

The plants, which arise from a basal disc, are filamentous with irregular pinnate branching from the main axes. The lower portion becomes multi-seriate by sept at ion and then resembles *Sphacelaria* although the growth is always interring calary. The sexual plants develop intercalary, tubular, plurilocular organs (gametangia) which are produced by the transformation of one or more cells of the main filament. The swarmers from these organs are thought to be male gametes.

Besides these organs there are larger, spherical, sessile, uninucleate oogonia partly immersed in the branches. These plants were formerly known as *Scaphospora speciosa.* The asexual plant reproduces by means of quadrinucleate spores formed singly in stalked or sessile mono-sporangia which are terminal or intercalary. Meiosis has been reported as occurring in these sporangia and this would be expected if they were primitive tetrasporangia. It seems therefore that the plants known as *Haplospora globosa* and *Scaphospora speciosa* are simply alternate phases of one and the same species, the latter being the sexual generation.

SPHACELARIACEAE: SPHACELARIA

The plants grow attached to stones or other algae by means of basal discs or rhizoids that have spread down from the lower cells of the axis. The basal disc may give rise to horizontal stolons and so enable the plant to spread.

The erect system is filamentous and the type of branching is termed hemi-blastic in which the laterals, whether of definite or indefinite growth, arise from the whole height of a cell representing the upper segment derived from the division of a primary segment. If the branching is pinnate practically every superior segment produces branch initials, though in some cases the initials may remain dormant and the branching is then less luxuriant. Branches of limited growth bear colourless hairs which are cut off as small cells to one side of the apical cell.

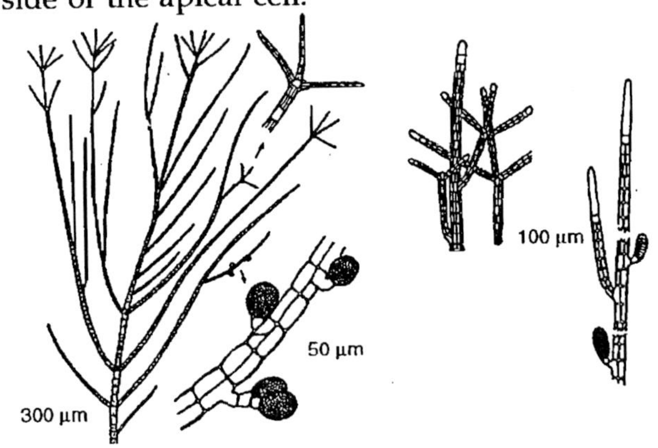

Fig. Sphacelaria

As the apical cell elongates the hair comes to occupy a lateral position. These hairs may disappear with age. This type of hair formation is regarded as sympodial, the hair initial being the true apical cell while the so-called branch axis is regarded as a subsidiary branch. Branching, which occurs in this way is said to be holo-blastic.

Growth in length takes place by transverse divisions of the apical cell, and when the basal cell so cut off has reached a certain length it divides into a superior and inferior segment.

These segments exhibit no further enlargement, though they divide transversely and longitudinally to form the corticated thallus characteristic of the genus. This type of vegetative system is said to be leptocaulous, and the filament diameters are essentially uniform throughout.

Vegetative reproduction takes place in this genus by means of modified branches or propagules which vary in shape from wedge-like to di- or triradiate with long or short pedicels. Unilocular and plurilocular structures are borne terminally or laterally on simple or branched laterals and are either sessile or shortly pedicellate. The life cycle of *Sphacelaria bipinnata* has been worked out in some detail and is probably typical of the genus.

The asexual generation bears both uni- and plurilocular sporangia. Zooids from the former are said to fuse in clumps, but those from the latter germinate directly and represent a means of accessory reproduction. The cytology of the clumps has not been studied though meiosis occurs in the unilocular sporangia. These clumps must give rise to the isomorphic sexual generation which reproduces by means of isogametes liberated from plurilocular gametangia. *S. harveyana* and *S. hystrix* are anisogamous. Male and female gametangia are born on separate plants in *S. harveyana* while in S. *hystrix* they can occur on the same plant.

CLADOSTEPHACEAE: CLADOSTEPHUS

The plants, which are bushy in appearance, arise from well-developed holdfasts and are characterized by ecorticate branches arranged in whorls with tufts of hairs just below their apices. Cells just below the apex divide to give a number of branch segments, this type of branching being known as polyblastic.

The first segment of these laterals of limited growth cuts off a cell or cells that form a cushion of tissue belonging to the main axis. Later the spaces between laterals may be filled by tissue so that the bases are completely enveloped. As a result of this septation the older branches, although hemi-blastic in origin, appear to arise only from the tops of the original superior segments. The lower portions of the segments may later give rise to secondary whorls which are said to arise meroblastically.

The main axis is corticate because both upper and lower primary segments undergo abundant division. In addition the cells so formed enlarge in width and length so that the axis increases in size from the apex downwards. Such a form of growth is termed auxocaulous. *Cladostephus* also forms long branches which arise in a hemi-blastic manner.

Both unilocular sporangia and plurilocular gametangia are formed on separate plants on special branchlets which arise from the rhizoidal cortex in the internodes between the whorls of vegetative branches.

STYPOCAULACEAE: HALOPTERIS

The pinnate frond arises from a distinct basal pad, the plants in summer having the appearance of shaggy tufts, while in winter the branching appears more regular as surplus branchlets are shed. The inner cortex of the central axis is composed of a number of cubical cells and there is also an outer cortex of rhizoidal cells (pericysts), the whole forming a pseudo-parenchyma.

Because there is no enlargement of the original cells, *Halopteris* is lepto-caulous. Pericysts are only found in the main axis and in branches of unlimited growth. Branching in *Halopteris* is holoblastic. At an early stage the branch initial cuts off a small upper axillary cell. This cell is regarded as the true apical cell so that the branch is really a lateral of a diminutive branch axis. In some species the axillary cell may later give rise to a tuft of hairs. Vegetative reproduction can take place by fragmentation, the fragments developing rhizoids and a new axis.

Sexual reproduction is by means of antheridia and oogonia and asexual by unilocular sporangia and zoospores. The reproductive organs are borne in the axils of laterals, when they must be regarded as terminal on axillary shoots, or on branchlets arising from pads derived from the axial initial. In *H. scoparfa* the sori are united into spikelets, the lower laterals being sterile and only the upper being fertile. A number of species are known, especially from the South Pacific, but their identification is not always easy.

CUTLERIACEAE: CUTLERIA

The gametophyte and sporophyte generations differ in their seasonal occurrence, the former being a summer annual while the latter is a perennial reaching its maximum vegetative phase in the Northern hemisphere in October and November with a peak fruiting period in March and April. The gametophyte is an erect, flattened thallus with irregular branching brought about by periodic failure of the marginal filaments to fuse together. In cross-section the plant consists of larger cells in the centre and smaller ones in the epidermis.

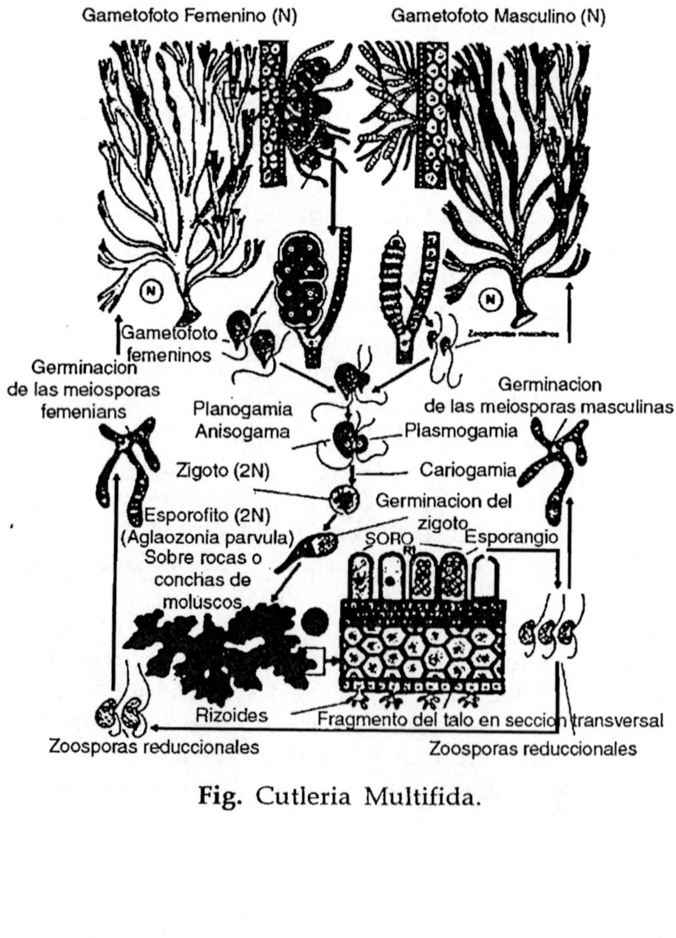

Fig. Cutleria Multifida.

The thallus and apices are clothed with tufts of hairs, each with a basal growing region, while the female gametangia and antheridia, which are borne on separate plants, occur in sori on both sides of the thallus. The antheridia are borne normally on much-branched threads. The mature plurilocular antheridium contains about 200 antherozoids that are much smaller than the mature female gametes.

The female gametangia, with which hairs are sometimes associated, are also formed from superficial cells that divide into a stalk cell and a game-tangium initial. The ripe gametangium contains fewer female gametes which, after liberation, remain motile for a period of from five minutes to two hours, while the antherozoids can remain active for about 20 hours. No fertilization occurs while the female gametes are motile. After settling they round off and attract large numbers of antherozoids chemotactically though this power is lost within 3-4 hours.

Upon germination a small columnar structure is first formed and then a flat basal expansion grows out from its base to form the adult sporophyte, which is a prostrate expanded thallus attached by means of rhizoids. It is morphologically different to the gametophyte and was initially thought to be a separate genus, *Aglaozonia.* The female gametes may germinate partheno-genetically to give sterile haploid *Aglaozonia* plants.

The sporophytic thallus is composed of large cells in the centre with superficial layers of small cells. The sessile unilocular sporangia borne on stalk cells and sometimes accompanied by deciduous hairs are clustered in palisade-like sori or are scattered irregularly on the upper surface of the thallus. Eight to thirty-two haploid zoospores are formed in each sporangium. The zoospores on germination give rise to *Cutleria* plants.

DICTYOTACEAE: DICTYOTA

In *D. dichotoma* the flattened thallus exhibits what is practically a perfect dichotomy because there is always a median septation of the apical cell. In transverse section the thallus is seen to be composed of three layers, a central one of large cells and an upper and lower epidermis of small assimilatory cells from which groups of mucilage hairs arise.

The sex structures are borne in sori on separate plants, the male sorus being composed of as many as 300 plurilocular antheridia surrounded by an outer zone of sterile cells. At the formation of an antheridium a superficial cell divides into a stalk cell and an antheridium initial, which then partitions into individual antheridial cells. The mature antherozoid is pear-shaped with only one flagellum, though two basal bodies can be seen, one of them being vestigial. Fewer ova are produced and it has been estimated that there are about 6,000 antherozoids available for each ovum.

In the oogonial sorus the large, fertile oogonia, 25-50 in number, are situated in the centre and surrounded by sterile cells on the outside. The oogonia likewise arise from superficial cells that divide into a stalk cell and oogonium initial, and each oogonium produces one ovum. Fertilization occurs in the water, and the eggs may be chemotactic.

Unfertilized ova may develop parthenogenetically; such plants, however, always die in culture, though it is possible that in nature they may persist. The sex organs are produced in regular crops/the new sori appearing between the scars of the old, and when the whole of the surface has been used up the plant dies.

The zygote develops into a morphologically similar plant which reproduces by means of spores that are normally formed in tetrads in superficial sporangia. At sporangium formation an epidermal cell swells up in all directions, and after a stalk cell has been cut off the sporangium initial divides meiotically to give the four tetraspores. In some cases the contents of a tetrasporangium fail to divide into four spores but germinate as a whole.

In other cases, such as in *Zonaria* and sometimes in *Pocockiella,* eight spores may be produced instead of the usual four. While the sex organs are produced in rhythmic crops there is no such periodicity in the case of tetraspores.

Three kinds of rhythmic periodicity for the sex organs of *Dictyota* have been described from different localities:

1. In Wales the sori require 10 to 13 days to develop while in Naples 15 or 16 days are necessary, the gametes being liberated about once a fortnight in both areas.

2. In North Carolina liberations occur once a month, at the alternate spring tidal cycles, although only eight days are required for the development of the sex organs.
3. In Jamaica the successive crops take a very long time to mature,e.g. very little change can be seen even after 22 days.

This results in almost continuous fruiting with two successive crops overlapping. There is obviously a considerable difference in the behaviour of species of *Dictyota.*

Wherever the plants occur, the bulk of the gametes are usually liberated in a single hour around daybreak. It has been suggested that light is the determining factor during intertidal periods; when plants are removed to the laboratory however, the periodicity is maintained.

The phenomenon is not confined to *Dictyota* because regular or irregular periodic liberations have been recorded for species of *Halicystis, Sargassum, Cysto-phyllum, Padina, Nemoderma* and *Ectocarpus.* Light does have some effect since under laboratory conditions a few hours of artificial light at approximately lunar intensity at any period of the night are sufficient to produce the rhythm. It has been observed that the plants in North Carolina always fruit at the time of full moon.

CHORDARIACEAE: MESOGLOIA

In this genus there is a single central strand terminating in a hair and having a distinct intercalary meristem just below the apex. The cortex is formed of short lateral filaments with somewhat globose, terminal cells that are packed in a gelatinous material.

The hairs, which are frequently worn away in the older parts of the thallus, occupy a lateral position, but owing to inequalities of growth they may appear to be terminal. The unilocular sporangia are ovoid and are borne at the base of the cortical filaments, but the elongate plurilocular sporangia, which are only known for *M. levillei,* replace the terminal portion of the assimilatory hairs and are therefore always stalked. Meiosis takes place in the unilocular sporangia during zoospore formation, and the zooids from the unilocular sporangia germinate into a minute winter gametophyte that bears plurilocular gametangia.

The zooids from these gametangia fuse and the zygote develops into the characteristic basal disc from which the central erect filament of the macroscopic plant arises. There is thus an alteration of morphologically distinct generations in this species.

In *Eudesme* and *Myriogloia* the branched mucilaginous plants possess a dense growth of hair and differ from *Mesogloia* in having more than one central strand in the medulla; there are also developmental differences.

In the genus *Chordaria* development has preceded a stage farther and the branched cartilaginous fronds possess a firm, pseudo-parenchymatous medulla of closely packed cells that have become elongated in a longitudinal direction. The cortex is composed of crowded, radiating assimilatory filaments, which are either simple or branched, the whole being embedded in a thick layer of jelly which makes the plant slimy to touch.

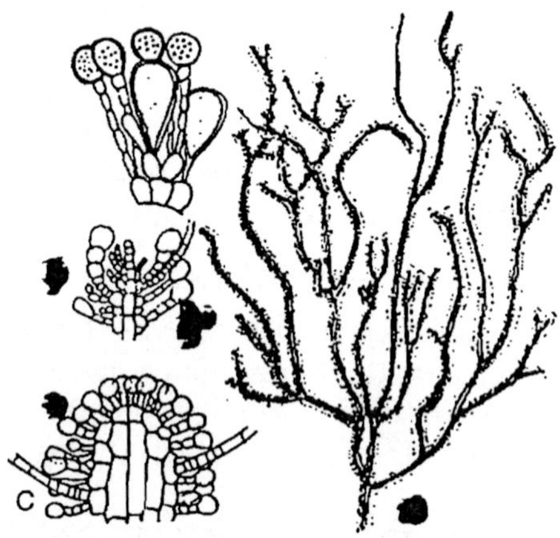

Fig. Chordaria Divaricata.

This type of structure, even though the growth is still confined to the apex, marks the highest development of the cable type of construction. Only unilocular sporangia are known.

In recent years culture work has led to the discovery of a dwarf generation,. In *C. flagelliformis* the dwarf generation produced plurilocular reproductive structures, which are probably asexual accessory bodies since no fusion of products was observed. Kornmann does not regard the dwarf generation as gametophytic.

It seems that the zoospores of the unilocular sporangia can either germinate into this dwarf thallus which diploid and filamentous or into a new diploid macroscopic plant. Which occurs appears to be dependent on sea temperature; at high temperatures the macroscopic plant develops and at low temperatures the dwarf filament.

ACROTRICHACEAE: ACROTHRIX

Species of this genus are found in the northern hemisphere. The plants arise from a disc-like holdfast and are basically uniaxial in construction, the main axial filament being terminated by a hair. Growth in length is normally trichothallic though the hair may later be lost leaving a pseudo-apical meristem. Cells cut off from the meristem enlarge and produce whorls of primary laterals of limited growth.

The basal cells of these laterals increase in size and give rise to short perimedullary threads that form the cortex which in older parts of the thallus may separate from the axial filament. As this latter undergoes great alteration or disintegrates the thallus can become hollow. Only unilocular sporangia are known and they are borne at the base of the assimilators. Except for the intercalary growth and terminal hair *Acrothrix* shows considerable resemblance to *Spermatochnus.*

CORYNOPHLAEACEAE: LEATHESIA

This genus provides an example of modification in the cable type of construction. The young plant arises from a small, creeping, rhizomatous portion and is composed of a close mass of radiating, dichotomously branched filaments which are sufficiently compacted to make the plant mass solid.

From these medullary filaments there arises a cortex of densely packed assimilatory filaments from which occasional colourless hairs emerge. The young plants are subspherical at first, but with increasing age the central medullary filaments commence to disintegrate and as a result the mature thallus becomes hollow and irregularly lobed. Plurilocular and unilocular sporangia are known, the zoospores from the ovoid unilocular sporangia germinating to disc-like plantlets on which plurilocular structures ultimately appear.

These represent accessory reproductive bodies since the swarmers give rise to other similar plantlets. A new adult plant arises as a vegetative bud so that the dwarf plants must be diploid. No dwarf gametophytes have been reported. The zoospores from the plurilocular sporangia of the macroscopic plant produce dwarf plants that can reproduce by means of plurilocular sporangia or can give rise to buds which develope into new adult plants.

ELACHISTACEAE: ELACHISTA

Juvenile plants possess a horizontal portion from which a number of erect filaments arise, so that in the early stage *Elachista* is comparable morphologically to *Eudesme.* In the older plant the erect filaments develop to form a cushion composed of densely branched filaments matted together and only becoming free at the surface.

The mature thallus comprises three distinct regions. In the dense central portion there is a mat of profusely branched threads which gives way to a looser peripheral zone of assimilatory filaments (sometimes termed paraphyses). Outside this is an outermost zone of free, long assimilatory filaments with chloroplasts. Morphologically it is therefore a reduced and modified *Eudesme* structure.

The species are epiphytic on other algae, the genus being represented in both hemispheres. In most species unilocular sporangia arise from the base of the assimilators, the distal portions of the latter being modified to form plurilocular sporangia. The zooids from the unilocular sporangia germinate in late autumn to give a branched, thread-like, microscopic gametophyte which persists throughout the winter. In late winter and spring plurilocular gametangia develop on the minute gametophytes, and when the gametes have been liberated fusion occurs and the zygote germinates on the host into a new macroscopic *Elachista* plant.

Exceptions to the normal type of life cycle are found in *E. fucicola* and *E. stellaris.* Only unilocular sporangia are known and the zoospores develop into plantlets that give rise to new sporophytes as lateral outgrowths. Plurilocular sporangia have recently been described in the creeping basal layer of *E.fucicolay* but they are probably accessory organs.

MYRIONEMATACEAE: MYRIONEMA

This represents one of the highly reduced genera, but the fundamental basic cable construction of prostrate thallus and erect threads can be found, though the erect threads do not branch. The genus is of wide distribution, the commonest species being *M. strangulans* epiphytic on *Viva* and *Enteromorpha.* The various species form thin expansions or minute flattened cushions or discs that are very variable in shape and from which numerous, closely packed, erect filaments and hairs arise. The basal monostromatic portion of the thallus, which has a marginal growing region, is composed of crowded, radiating filaments that may, on rare occasions, penetrate the host plant.

The unilocular sporangia, which are not borne on the same plants with pluri-locular sporangia, give rise to haploid zoospores, and these develop into a thread-like gametophytic plant bearing long filaments. The gametophyte phase is absent in M. *orbiculare* and *M. magnusii.* Although *M. strangulans* and *M '. Feldmanni* possess a life cycle typical of the Chordariales, it has been suggested that some of the other species placed in the genus or in allied genera may have isomorphic alternation, in which case they may need to be removed to the Ectocarpales.

SPERMATOCHNACEAE: SPERMATOCHNUS

This is essentially a corticated type. The filamentous, cylindrical, branched thallus is derived from a single central axis with a definite apical cell. Each individual cell of this axis filament segments at one end and so definite nodes are formed.

The corticating filaments arise from the nodes, and growth of the cortex is secured by tangential division of the primary corticating cells, though later more filaments may grow on top of them. The outermost layer of the cortex bears the assimilatory filaments and hairs. As the plants become older mucilage develops internally and forces the cortex away from the primary central filament, although a connection is maintained by the threads from each node.

Unilocular sporangia, together with clavate paraphyses, develop in sori, the sporangia arising from the base of the sterile threads. The life cycle has not yet been worked out, but if it is at all comparable with the other closely related genera then the zoospores should give rise to a microscopic plant phase or sexual generation.

STILOPHORACEAE: STILOPHORA

The plants are bushy and attached by a holdfast disc. Structurally they are multiaxial with growth taking place by divisions of the apical cells of the axial filaments. This condition has probably arisen through the loss of trichothallic growth. A parenchymatous cortical envelope is formed as in *Spermatochnus,* the peripheral cells of this cortex dividing to form a small-celled superficial layer that bears colourless phaeophycean hairs and simple or branched assimilators arranged in localized areas.

The lower cells of the assimilators give rise to unilocular or plurilocular (accessory) sporangia that appear macroscopically as wart-like sori. Zoospores from the unilocular sporangia germinate to a dwarf ectocarpoid gametophyte that reproduces by means of plurilocular gametangia. Zoospores from the plurilocular sporangia give rise to cushion-like protonemal thalli that bear unilocular and plurilocular sporangia.

SPLACHNIDIACEAE: SPLACHNIDIUM

This is a monotypic genus of considerable phylogenetic interest. The single species, *S. rugosum,* in found in South America, South Australia, South Africa and New Zealand. The plants occur scattered intertidally on the rocks in the lower half of the littoral zone. They are tubular, sparsely branched and contain a large amount of mucilage. Structurally the plant is multiaxial but the axes become widely separated and only at the periphery do the lateral branches proliferate to form a pseudo-parenchymatous cortex.

At the apex of the axis or a branch there is sometimes a cell of the green endophyte *Codiolum kuckuckii,* and this was formerly regarded as the apical cell. The apical parts are clothed with short, clavate assimilators that later become shed so that the surface is then formed by a compact group of small cells.

Reproduction takes place by means of unilocular sporangia that develop in sori with tufts of hairs in shallow conceptacles. The zoospores after liberation swim about for a short time and then settle on a substrate where they germinate into a dwarf, branched, filamentous, ectocarpoid generation that reproduces by means of plurilocular organs; this is probably the gametophyte.

NOTHEIACEAE: NOTHEIA

The filiform sporophyte of the sole species, *N. anomala*, grows out para-sitically from the base of conceptacles of the Fucalean *Hormosira banksii* and from its own conceptacles. There is an apical growing zone of three cells that give rise to filamentous rows of cells. Later divisions in different planes lead to a pseudo-parenchymatous thallus in which epidermis, cortex and medullary tissue can be distinguished. The apical filamentous development is characteristically Chordalian. True branching is rare, such branches being recognized by the lack of any basal constriction.

Most of the apparent branches, which are constricted at the base, represent new plants that have developed from zygotes produced on a filamentous gametophyte that grows inside the conceptacle. In those portions of *Hormosira* that are attacked by the parasite the hollow of the vesicle-like internode becomes filled up by new tissue formed as a result of the stimulation, but the parasite is apparently unable to attack *Hormosira* unless the host is growing in areas where it is continuously submerged.

The fertile conceptacles contain either mega- or microsporangia, the former being much more frequent. Each megasporangium gives rise to eight pyriform zooids (originally believed to be non-motile ova) that germinate apparently in the conceptacle to a branched filamentous gametophyte with erect colourless hairs. Cells of erect branches can each produce a swarmer which is presumably a female gamete.

It seems likely from the construction of the thallus that

some of these germinate inside the conceptacle to give a new macroscopic plant. Others must escape and infect *Hormosira* plants. The microsporangia each give rise to 64 zooids, the fate of which is not known. They could presumably either develop to a dwarf-male gametophyte or else function as a male gamete. These features are of special importance in considering the origin of the Fucales.

SPOROCHNACEAE: SPOROCHNUS

The plants are moderately large and grow by means of a horizontal meristem that lies beneath the terminal tuft of assimilatory hairs. Young plants consist of a simple erect thread anchored by rhizoids. The lower part represents a pedicel which plays no further part in development. The upper part forms the primary hair and the cell between the upper and lower parts is the archi-meristem from which the adult thallus is produced by longitudinal divisions in various planes.

Threads which grow downwards are produced and coalesce to give the pseudoparenchymatous body. The sporangia are borne laterally on special, branched, fertile threads in the soral region. The gametophytes are ectocarpoid with small antheridia and swollen terminal cells that are believed to be oogonia.

DESMARESTIACEAE: DESMARESTIA

The plants are bushy and usually of some size, especially the Pacific species. They sometimes bear swellings (galls) which are caused by a copepod. The erect, cylindrical or compressed thallus arises from a disc-like holdfast and exhibits regular pinnate branching, often with a distinct mid-rib.

The thallus is composed of a single prominent central row of large cells with an intercalary meristem at the base of the terminal hair. The large central cells, the transverse walls of which are pitted, are surrounded by cortical cells, the primary cortex being derived from the basal cells of the primary laterals. These divide periclinally to give a many-layered cortex, the cells of which are smaller toward the periphery. In the older parts, small hyphal cells are interspersed in the cortex; these originate as a result of secondary activity from the cortical cells.

The sporophytic plants have only unilocular sporangia and these are slightly modified cortical cells. The zoospores escape in a mass and germinate to produce dioecious filamentous gametophytes which are heteromorphic. The smaller male plants produce terminal antheridia from each of which is liberated a single antherozoid; the larger female plants produce swollen oogonia.

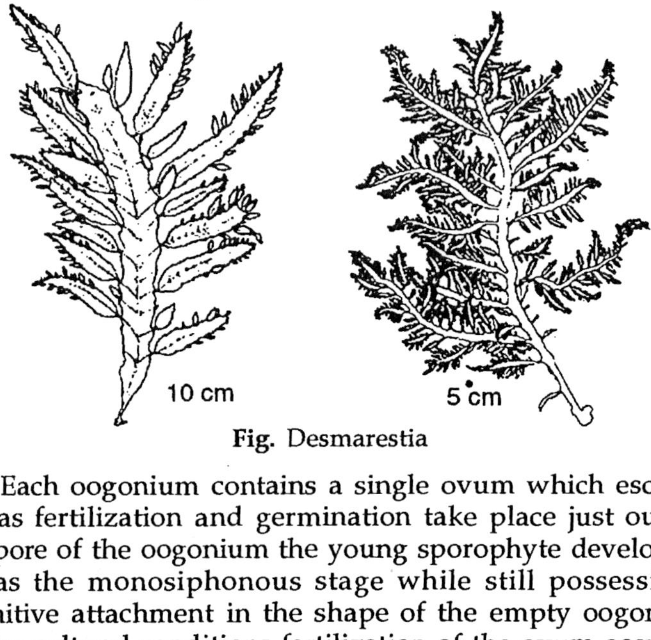

Fig. Desmarestia

Each oogonium contains a single ovum which escapes, but as fertilization and germination take place just outside the pore of the oogonium the young sporophyte develops as far as the monosiphonous stage while still possessing a primitive attachment in the shape of the empty oogonium. Under cultural conditions fertilization of the ovum occurs in the oogonium, the new sporophyte commencing life attached to the gametophyte. Cortication, which is best observed near the apex of old plants, commences in the young plants after a few weeks.

MYRIOTRICHIACEAE: MYRIOTRICHIA

The minute epiphytic species of the genus *Myriotrichia* are the simplest plants in the Dictyosiphonales. The main axis bears the sporangia, hairs and assimilators, though the last-named are apparently not always present. Uni-and plurilocular structures generally occur on different individuals and this suggests alternation of isomorphic generations, but there is not sufficient cytological evidence to support this view.

STRI ARIACEAE: STRIARIA

Striaria attenuata is a widespread sublittoral epiphyte with a branched cylindrical thallus. The branches are usually opposite and taper to their apex and base but terminate in a hair. The thallus can be regarded as morphologically more advanced than that of *Myriotrichia*. In the old thallus a central cavity arises and this is surrounded by two layers of cells. The plants bear unilocular sporangia associated with unicellular paraphyses and phaeo-phycean hairs in bands to give the characteristic striations. According to Caram two types of life history occur in *Striaria attenuata*.

In one case there is an alternation between the normal thallus (with unilocular sporangia) and a filamentous gametophyte with plurilocular gametangia. The life history is therefore of the dimorphic diplohaplontic type.

In the other type of alternation the swarmers from the uni-locular sporangia may fuse and develop again into the sporophyte, as does a monomorphic diplont.

PUNCTARIACEAE: PETALONIA

The fronds are unbranched, expanded, membranous, leaf-like structures with an internal medulla composed of large, colourless cells interspersed with hyphae, and an outer layer of small, superficial, assimilatory cells. Unilocular sporangia are not known, but plurilocular sporangia arise from surface cells and produce swarmers that germinate to give a creeping basal thallus from which a new plant arises.

Chen have extended these observations and shown that *Petalonia* swarmers give rise to *Ralfsia clavata* whose zoospores in turn give rise to *Petalonia*. Wynne has indicated that *Ralfsia californica* may also be the alternate generation of *Petalonia fascia*. Light and temperature may determine the life history.

The creeping basal thallus can be branched or *ralfsia-like m* appearance. In this genus and the related, hollow, tubular genus *Scytosiphon* it seems that two kinds of dwarf plants can be produced. Some of these dwarf thalli reproduce by means of unilocular sporangia. These are presumably diploid plethysmothalli. This means that the macroscopic plant may be haploid. In *Scytosiphon lomentaria* the life history appears to be different in Japanese and European waters and it is possible there are really two species involved.

The only other group of brown algae with a similar morphological life history is the Cutleriales. Until fusion of swarmers has been seen it could only be assumption that these *'ralfsid** plants are sporophytes. It may be that *Petalonia* and *Scytosiphon* need to be removed to a separate order. In one instance, at least, the *'ralfsici* stage appears to give rise directly to the *Scytosiphon* form.

Other dwarf thalli reproduce by means of plurilocular organs which could be gametangia, though fusion of swarmers has not been seen. Further work on the life history of these algae is required in order to determine whether reduction division does occur.

DICTYOSIPHONACEAE: DICTYOSIPHON

The plants arise from small lobed discs and have either a few or many branches, the younger ones commonly being clothed with delicate hairs. In young plants there is the usual apical hair with a basal meristem. Later the hair is shed leaving an apical growing cell.

Subsequently a central medulla with four rows of large elongated cells develops and this may be penetrated by hyphae from the smaller cortical cells; in old plants it is often ruptured and the axis becomes partially hollow. Only unilocular sporangia occur, and the zoospores germinate to form microscopic plantlets; these represent the gametophytic generation and reproduce by means of plurilocular gametangia.

The gametes either develop parthenogenetically into a new plantlet or else two gametes from different gametangia, form a zygote which develops into a small ectocarpoid plant. This may either reproduce itself by means of pluri-locular sporangia or else gives rise directly to the adult sporophyte. In *D. chordaria* meiosis appears to have been suppressed as the unilocular zooids give rise to a new sporophyte.

ASPEROCOCCACEAE: ASPEROCOCCUS

The structure of the adult plant is essentially the same as

that of the two preceding genera except that the central filaments degenerate and the centre becomes filled with a gas. The fronds are simple or branched and bear small superficial cells with sporangia and mucilage hairs scattered over the surface in sori. Plurilocular and unilocular sporangia occur on the same or on different plants, the sori with unilocular sporangia containing sterile paraphyses.

The life cycle is simple, the zoospores from the unilocular sporangia germinating directly into a dwarf phase; this can reproduce itself by swarmers from uni- and plurilocular sporangia until the advent of favourable conditions enables the development of the macroscopic phase to take place once more. Only the diploid generation is therefore known. In *A. fistulosus* meiosis occurs in the unilocular sporangia.

If the swarmers fuse, the zygote develops into a *'streblonema phase,* so named after the brown alga it resembles, from which a new adult plant can arise. In this case there is no evidence for the existence of a gametophytic generation, nor has any evidence been obtained to show that such streblonemoid plants can reproduce themselves.

If no fusion of the swarmers takes place in *A.fistulosusandA. Echinatus,* the *"streblonemcf* phase is produced parthenogenetically, but under these circumstances plurilocular sporangia are formed which give rise to a new *'streblonemd** generation. No investigator, under such conditions, has however succeeded in obtaining macroscopic plants again and so it has been suggested that sex has been inhibited in these plants.

In *A. bullosus* zooids from plurilocular sporangia on the macroscopic thallus do not fuse but germinate directly to give rise to a series of plantules bearing plurilocular sporangia: these tide over the winter season, and then in spring young *Asperococcus* plants develop in place of the sporangia on the ectocarpoid plantules.

In the unilocular sporangia meiosis takes place and zooids develop into minute gametophytic plants that produce plurilocular gametangia. The gametes produced can fuse to produce a zygote that develops into a plantule from which a new macroscopic plant arises, but if there is no fusion then they merely develop into a new gametophytic generation.

Sauvageau also reported that the zooids from the unilocular sporangia may give rise to creeping filaments which later produce young plantules'as *Asperococcus*. This direct reproduction of the macroscopic plants can only be explained by work on the cytology of the different generations.

It will be evident that direct alternation of generations is obscured through the number of possible independent circuits and short-cuts. In *Colpomenia peregrina*, a member of a closely allied genus, there are two morphologically similar generations. The dioecious gametophytes appear in spring and reproduce by means of anisogametes that are formed in dissimilar plurilocular gametangia.

The zygote gives rise to a new adult sporophyte that reproduces in autumn by means of plurilocular sporangia. This life cycle is somewhat different from the others that have been described and a re-investigation would seem to be desirable, especially in view of the absence of unilocular sporangia, which is the usual place of meiosis in brown algae.

9
Fossil Algae

Many algal fossils are known and it is therefore only possible here to outline the main types of structure represented among them. The principal geological strata and the generally accepted time scale.

Because the algae are generally composed of 'soft tissue', preservation, particularly in the lower geological strata, either has not occurred or else has not been very good. Those algae that characteristically deposit lime, (e.g. Codiales, Dasycladales, Corallinaceae) or silica, (e.g. diatoms) form the groups that have been most successfully preserved and therefore about which most is known. There is little doubt that even at the present time fossilization of calcareous and siliceous algae is taking place, e.g. in the Arctic benthos where crustaceous Corallinaceae are particularly abundant.

In recent years more attention has been paid to Archeozoic or pre-Cambrian rocks and the result has been the discovery of fossil organic remains of a much greater age than previously reported.

CYANOPHYCEAE

Ontario in rocks that are two billion years old. *Animikiea septata* would appear to have been a member of the Oscillatoria-ceae. *Entosphaeroides amplus* with internal spores or endogonidia could be allied to *Chamaesiphon* or may have been bacterial in nature. Spore-like bodies put in the genus *Huroniospora* could have allies in either the Chroococcaceae or the Dinoflagellates. In addition there are three genera of uncertain affinities, *Archaeorestis, Eoastrion* and *Eosphaera.*

Other very early recognizable fossils are those described from late pre-Cambrian rocks (700-900 million years of age) near Alice Springs in Australia. The specimens tentatively ascribed to the Cyanophyceae are small septate filaments that taper from a maximum middle zone to the ends and were possibly enclosed in a sheath very like the modern *Phormidiwn* and *Lyngbya* species. All the above together with the following are best placed in a group called the Protophyceae.

SPONGIOSTROMATA

Much doubt has been thrown upon the authenticity of this group, some writers regarding them as structures which originated as diffusion rings ('liesegang' phenomena) in colloidal materials or perhaps in calcareous muds.

In the original description Walcott suggested an affinity to the Cyanophyceae, but as later workers could only distinguish a purely mineral structure they suggested the idea of diffusion phenomena. Discoveries of comparable algal concretions and laminations in the Bahamas, however, have made it extremely probable that these structures were formed by algae that collected and bound the sediment.

POROSTROMATA

The comments concerning the taxonomic placement of the Spongio-stromata also apply here. The Porostromata includes all fossil algae of unknown position whose structure consists of a collection of well-defined tubes.

The most widely represented genera are *Girvanella* and *Rothpletzella* which date from the Silurian and Devonian of Europe, U.S.S.R., Australia and Canada. The former is known from the Cambrian to Cretaceous but was most abundant in the Carboniferous. The thallus consisted of a loose mass of small twisted tubes that compacted to form a rounded or bean-shaped mass.

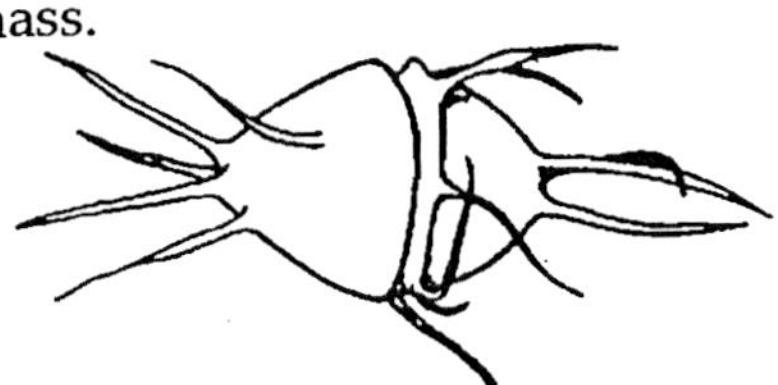

Fig. Hystrichodinium Oligacanthwn
(after Deflandre and Cookson)

They were possibly formed in much the same way as the

algal water-biscuits now found in South Australia. These range from tiny particles to thick bun-like forms 20 cm in diameter, whilst in them are to be found the tube-like remains of living species of *Gloeocapsa* and *Schizothrix*.

DINOPHYCEAE

A large number of fossil Dinoflagellates have been recorded. The earliest are reported from the Jurassic, but even then many of the fossil forms could be ascribed to such extant genera as *Gonyaulax, Gymnodinium, Peridinium,* and *Ceratium.* Among extinct genera in the Gymnodinioideae are such forms as *Diconodiniwn* from the Australian Cretaceous. The genus showed the characteristic girdle, and the epitheca terminated at the apex in a short stumpy projection.

Among the Peridinioideae, extinct genera are represented from the Jurassic to Tertiary by such forms as *Palaeoperidiniwn.* A fossil member of the Gonyaulacaceae is the Cretaceous *Rhynchodiniopsis,* while *Raphiododinium* is peridinioid with long spiny protuberances. All of these forms, like many other of the fossil genera, possess the characteristic dinoflagellate orientation with a distinct girdle, and in many cases there is a sulcus or sulcal region.

Apart from those members that can be ascribed to extinct or extant families within the Peridinioideae or Gymnodinioideae there are some forms that are ascribed to families of uncertain position, e.g. *Hystrichodinium, Peridinites.* Again there is no doubt that these are dinoflagellates because of their quite distinct dinoflagellate orientation.

Associated with these planktonic algae but with a more ancient history, since they extend down at least into the Cambrian, are the wholly fossil Hystrichosphaeridae. The exact relationships of this group are not clear but they can at least be classed as Protists, and in some cases seem to have dinoflagellate organization.

By far the most common genus is *Hystrichosphaeridium* with representatives from the Silurian to the Tertiary. In *Micrhystridium,* known from the Silurian to the Tertiary, the characteristic spiny projections are not nearly so ornate or complex. In Kansas there is evidence of two successive Cretaceous blooms which consisted of a dinoflagellate hystrichosphaerid phytoplankton. Such a bloom must have involved adequate supplies of nutrients, including Vitamin B_{12}, and there must also have been certain optimum physical factors.

BACILLARIOPHYCEAE

Recognizable fossil diatoms are known from the Upper Jurassic, but they are most abundant in the late Cretaceous and Tertiary. The majority can be ascribed to present day forms. Fossil diatom deposits are known as Kiesel-guhr and are valuable commercially.

Freshwater diatom deposits, especially those from the late-glacial and postglacial, provide a means of deducing changes that may have occurred in lakes and drainage areas. Round has demonstrated how variation in diatom species and frequency can lead to interpretation of environmental changes. Thus the dominants in the early period, e.g.

Gyrosigma, Cyma-topleura, indicated waters of a higher alkalinity than at present. At the time of the Boreal-Atlantic transition, the early alkaline species began to be replaced by more acidic species, e.g. *Eunotia, Anomoeoneis,* which subsequently increased greatly. As Round points out this change is correlated with increasing warmth of the climate, with a decline of the Birch/Pine forest climax and the finish of further base supplies as the rock surfaces became leached.

Another example of the use of fossil diatoms for interpretive purposes comes from the Presto fjord. In this case, changes in the proportions of freshwater, brackish and marine diatoms showed that in the pollen analytical period Vila the Baltic was very slightly brackish, but in Vllb, after a period of reduced salt there was a rise in salinity. Finally, at the end of IXb, there was a recent reduction in salinity.

CHLOROPHYCEAE

It seems now that the oldest known representatives must be the large septate filaments of *Gunflintia* from the Gunflint chert of Ontario and described by Barghoorn and Tyler, and similar filaments described by Barghoorn and Schopf from the late Precambrian of Australia. These were multicellular and unbranched and thus very suggestive of the present day Ulothricales.

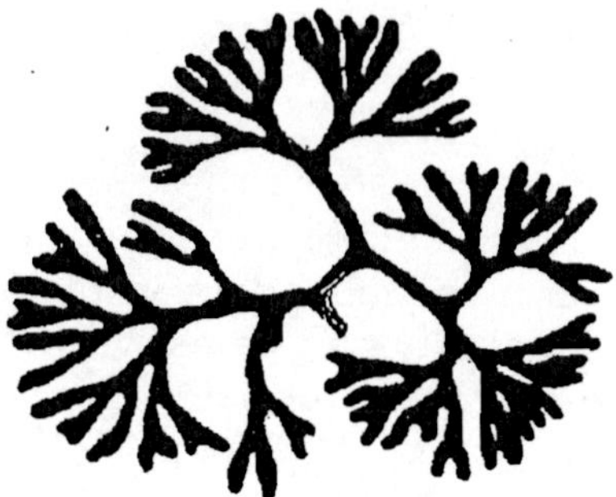

Fig. Codiaceae.

CODIACEAE

This was one of the most important families of rock-building algae, especially from the Ordovician to the Pennsylvanian and from the Jurassic to the present. Examples of these algae are *Boueina*, an unbranched form from the Lower Cretaceous, and *Palaeoporella*, which was composed of hollow cylinders of funnel-shaped bodies with slender forked branches, from the Lower Silurian. *Dimorphosiphon*, from the Ordovician, is generally regarded as the oldest-known member of the Codiaceae and has been tentatively related to *Halimeda*. It was about 10 mm long and composed of branched tubular cells without any cross walls, the cells being embedded in a calcareous matrix.

Genera that bear little resemblance to present day ones are the Devonian *Abacella* and *Lancicula* with their peculiar tufted branchlets. In the Permian the Codiaceae represented an important family of fossil algae with some nine genera.

DASYCLADACEAE

This group contains a very large number of the fossil algae. It reached its maximum development and abundance in the upper Jurassic and lower Cretaceous, and in those days was much more important than at present. Of the 58 known genera of the Dasycladaceae some 48 are extinct. *Cyclocrinus* grew to about 70 mm and looked like a miniature golf ball borne on the end of a stalk. Narrow branches arose at the apex of the stalk and each terminated in a flattened hexagonal head; as the edges of adjoining heads were fused together to form the outer membrane, which was only weakly calcified, the cell outlines were clearly visible.

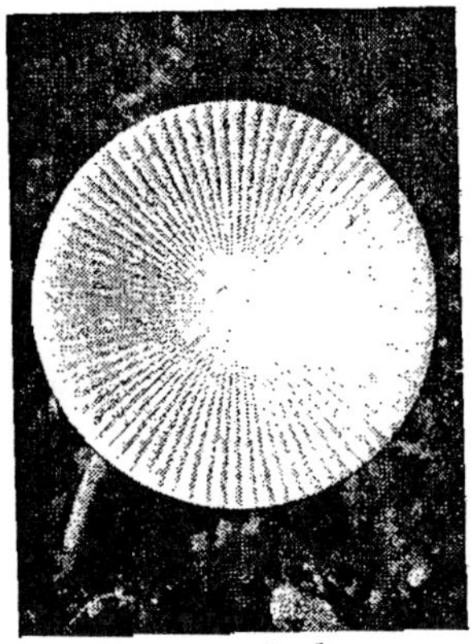

Fig. Dasycladaceae

Primicorallina, from the Ordovician, had a stem beset with radially arranged branches, each of which branched twice into four branchlets.

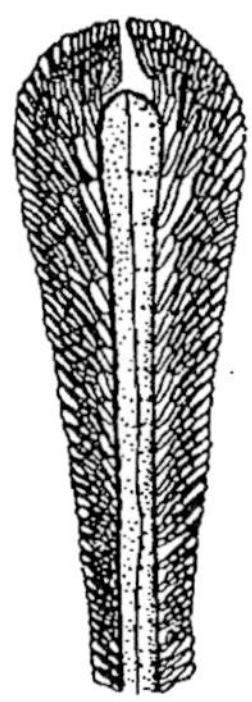

Fig. Dasycladaceae. Palaeodasycladus

Rhabdoporella seems to be one of the most primitive genera as it is represented by a purely cylindrical shell that is studded with pores through which the threads passed. It is known from the Ordovician and Silurian. The type of structure found in *Diplopora* was also shown by many other forms from the Middle Triassic of Eurasia.

It was a few centimetres long and bush-like in appearance,

the main stem being covered with whorls of branches that arose in groups of four, each bearing secondary branches that terminated in hairs. In the older thalli the outer part of the branch dropped off leaving a scar on the calcareous shell. The sporangia are reported to have been modified branches.

Palaeodasycladus, from the Lower Jurassic, bears a resemblance to the living species of *Dasycladus.* Near the base there were only primary branches, while higher up secondary and tertiary branches were to be found.

Lagynophora, a genus from the Lower Eocene, can be ascribed to this group, while *Palaeonitella,* from the Middle Devonian, may belong here also although its affinities are not so clear.

Charophytes are known from much older rocks and took the form of gyrogonites. The earliest of these is *Trochiliscus podolicus* from beds of the Lower Devonian. In all gyrogonites it is important to note that the remains comprise only the calcareous inner part of the oogonium with the enclosed oospore.

The coronal cells, which are so important taxonomically, are invariably missing so that the relationship of the remains to living genera cannot be determined. In the Permian of North America the organ genus *Catillochara* includes all gyrogonites with five sinistral spirals.

The brown algae, mainly because of their lack of calcification are very sparsely represented as fossils. If there were large forms comparable to the present-day *Nereocystis, Macrocystis* and *Lessonia,* preservation does not seem to have taken place.

A number of problematical fossil algae are known, especially from the Silurian and Devonian, and these may well have been members of this group. Thus some of the described species of *Buthotrephis,* which are bush-like with fasciculate, flattened dichotomising branches, look not unlike a modern Dictyotalean. *Chaetocladus* with its jointed cylindrical stems and whorls of hairs is more of a problem, though there are algae in the Chordariales and Sphacelariales with which it has at least a superficial resemblance. The Devonian genus *Drydenia,* with flattened lamina and filamentous holdfast is not unlike a laminarian. The flattened thalloid *Enfieldia* and

Hungerfordia could be of either Phaeophycean or Rhodophycean affinities.

Parker and Dawson have recently described some interesting fossils from Californian Miocene deposits. Amongst them was one genus *Jules-craneia*, which they considered as an intermediate between the two present-day Californian kelps, *Nereocystis* and *Pelagophycus*.

Also present were forms very similar to the genera *Cystoseira* and *Halidrys*, viz. *Palaeohalidrys* and *Cystoseirites* as well as algae, e.g. *Palaeocystophora*, having affinities to the present extant southern genus, *Cystophora*. This lastnamed example raises interesting queries as to why the living forms are now restricted to the southern oceans.

The lime-encrusted Melobesiae are represented from the Cretaceous upwards by species of *Archaeolithothamnion*, *Dermatolithon*, *Lithothamnion*, *Lithophyllum* and *Goniolithon*, some of them being distinguished only with difficulty from living forms. The Corallinaceae are also represented in the Palaeozoic by extinct members of present-day living genera. There are a large number of forms assigned to an extinct family, the Solenoporaceae, which existed from the Ordovician up to the Jurassic, and recently their structure and systematic position has been revised.

They formed nodules which ranged in diameter from the size of peas up to several centimetres and the cells of the nodules were arranged like those *ofaLithothamnion* although the cross walls were not well marked. In most forms the cells are distinctly larger in size than in most present-day *Lithothamnia*. A hypothallus, if present, appears to have been thin and poorly calcified. Conceptacles have been described in these plants and were presumably associated with rdproduction.

The so-called organs of reproduction in the Solenoporaceae have been the subject of much discussion. There appear to be three types including the wide dome-shaped structures regarded as conceptacles. The others are star-like groups of extra large cells or single, elongated, spindle-shaped, cylindrical or ovoid subjects. The three main Palaeozoic genera of this family are *Soleno-pora*, *Parachaetetes* and *Ungdarella*.

The relationships of the various fossil genera to the living ones have been postulated by Johnson. Other genera that have been ascribed to the Rhodophyceae are *Palmato-phycus* (Silurian) with feathery branches and fine hairs, *Thamnocladus* (Devonian) with a fleshy lamina transversed by a central axis or strand, *Foliophycus* and *Cuneiphycus.* In the Permian the extinct Gymnocodiaceae are regarded as having been very similar to species of *Galaxaura.*

In Californian Miocene deposits a few fleshy Rhodophyceae resembling extant members of the Cryptonemiales, Gigartinales and Ceramiales have recently been discovered. These include *Paleopikea, Chondrides* and *Delesserites.*

NEMATOPHYCEAE: NEMATOPHYTALES

Two genera are now grouped in this assemblage which was established by Lang, and although he regarded these forms as land plants, nevertheless they have so many features in common with the algae that it is felt proper to include them here, as does Johnson.

It is perhaps almost too speculative to suggest that they represent transmigrant forms, but it would appear they they must either be regarded as highly developed algae which adopted a land habitat, or else as the most primitive of all true land plants. The two genera agree closely in their morphological structure, and although they are both frequently found associated with each other in the Devonian rocks the two structures have not yet been found in organic connection.

In spite of this it is not unlikely that the leafy *Nematothallus* was the photosynthetic lamina of the stem-like *Nematophyton* and may also have functioned as the reproductive organ.

In the lowest strata the plants are found associated with remains of marine animals, which suggests that they grew under marine or brackish conditions, while in the higher strata they occur in beds, which are regarded as freshwater or continental, where they are associated with plants that were undoubtedly terrestrial. The presence of spores in *Nematothallus* has been regarded as rendering it unlikely that the plants were algal in nature, but the spores may be comparable to the hard-walled cysts such as are to be found in *Acetabularia.*

The genus *Nematophyton* is found in Silurian and Devonian rocks where it was first described under the name *of Prototaxites* and referred to the Taxa-ceae, but subsequently it was accepted as an alga and re-named *Nematophyton* or *Nematophycus.* Later the name *Prototaxites* was unfortunately revived and it was placed in the Phaeophyceae, while Krausel stated that it must have had the appearance *of Lessonia.* The valid name is therefore *Prototaxites,* but it would seem more satisfactory to retain the better name of *Nematophyton.* The largest specimen is a stem up to two feet in diameter, but whatever the size of the stem it is usually composed of two kinds of tubes, one large and the other small.

The large tubes have no cross partitions, but in some species they are interrupted in places by areas which are wholly occupied by small tubes that in other parts of the thallus simply take a sinuous course between the large tubes. The wide tubes, in the specimens described by Lang, show no markings indicative of definite thickening, though striations have been seen in specimens from other localities. Around the outside of the central tissue there is a cortex, or outer region, composed of the same tubes where they bend outwards towards the periphery and eventually stand at right angles to the surface. The outermost zone of all is apparently structureless and may well have been mucilaginous.

Nematothallus is composed of thin, flat, expanded incrustations of irregular shape and up to 65 mm long by 10 mm broad, and constructed of the wide and narrow tubes. The thallus is surrounded by a cuticular layer that exhibits a pseudo-cellular pattern, and which includes firm-walled spores of various sizes within the cuticle and among the peripheral tubes; in *N. pseudovasculosa* the spores have a cuticle and so the suggestion was made that these were land plants or parts of a land plant.

The side tubes, which have thin, pale brown walls, are translucent in appearance and have characteristic annular thickenings. The cuticle, which is apparently readily detached, possesses distinct cell outlines that were probably made by the ends of the wide tubes from the ordinary tissue where they became fused together at the periphery, as in the living genera *Udotea* and *Halimeda.* Another species, *Nematothallus radiata,* is more imperfectly known.

The members of this group are strongly reminiscent of the Laminariales and Fucales, and it is tempting to suppose that they represent land migrants from one of these groups. There is no evidence, however, of any parenchymatous structure, so that if they were indeed members of that group they would be allied to forms with a multiaxial type of construction. Problems that have to be solved are:

(1) The cuticularized spores; while no such spores with hard outer walls are known from the brown algae they are recorded from the Chlorophyceae, e.g. *Acetabttlaria.* The suggestion, however, that the spores may have developed in tetrads adds a further complication, at any rate so far as an algal ancestry is concerned, because the Dictyotales and tetrasporic Rhodophyceae do not show the state of differentiation found in these fossil plants.

(2) The presence of a deciduous cuticle. In this connection one or two Laminariales are known to shed cuticles during reproduction, and the present authors have found a deciduous cuticle on plants of *Hormosira,* a member of the Fucales.

It may be suggested that the plants perhaps had the appearance of a *Lessonia,* or even of a *Durvillea,* and a stem diameter of a metre does not preclude them from being algal in character because several of the large Pacific forms may have stipes of almost this size. It has also been suggested that these forms are related to the Codiaceae, especially *Udotea,* and in certain respects it is true that they have the structure of a siphonaceous plant.

Here again there are several phenomena that need to be explained:

(*a*) The presence of two sizes of tubes,
(*b*) The presence of a cuticle,
(c) The presence of cuticularized spores,
(*d*) The large size of stem.

The answer to the last problem has already been suggested, but cuticles in the Codiaceae have not been recorded, although it is possible to detect a structure something like a cuticle in *Halimeda;* nor have any species been reported that possess two distinct sizes of tubes, although gradations in size occur in both *Udotea* and *Halimeda*. It must be admitted that there are no living members of the Codiaceae with stems that approach anywhere near the size of those of *Nematophyton*.

This, however, is not an insuperable objection as the Nematophytales may bear the same relation to the living Codiaceae that the fossil Lepidodendrons bear to the living Lycopodiales. For the present, however, the problem must be left in the hope that further evidence will be found.

ECOLOGY OF SALT MARSHES AND MANGROVE SWAMPS

Little is known about the ecology of algae in salt marshes in comparison with the rocky coast. This neglect has probably been due to the fact that the algae are often microscopic and more difficult to determine taxonomically. In practice, however, a detailed study of any one area often produces the somewhat unexpected result of a quite extensive flora.

An investigation of any salt marsh area shows that the algal communities offer a somewhat different aspect to the algal communities of a rocky coast. Thus on a fairly low marsh the 'Autumn Cyanophyceae' appear in autumn and early winter, they disappear and are replaced in spring by the *Ulothrix* community, which in its turn is replaced during the summer months by *Enteromorpha*.

Furthermore, as the continual deposition of silt gradually increases the height of the ground level in relation to the tide, the submergences become fewer and the communities are replaced by others because of the modified conditions. As a result there is a definite dynamic succession of the different communities over a period of years. On a rock coast there is no succession in time and the succession in space is static.

The continual replacement of one community by another as the marsh increases in height provides changes that are more akin to those found in land habitats. A survey of the algal communities of some salt marshes is set out in

The reason for this probably depends on the very different types of salt marsh. The Irish marshes are composed of a form of marine peat, the marshes on the west coast of England have a large sand component in the soil, the marshes on the south coast bear a tall vegetation of *Spartina* growing in a very soft mud, while the east coast marshes bear a very mixed vegetation on a mud that tends to be clay-like.

In New England the marshes are formed of a marine peat while in New Zealand the soil is either muddy or sandy clay. In spite of this, however, the Sandy Chlorophyceae, Muddy Chlorophyceae, Gelatinous Cyanophyceae *Phaeococcus* socies, *Catenella-Bostrychia*^ and the *Fucus limicola* consocies all have a wide distribution, though they may not necessarily appear at the same relative levels on the different marshes. On the whole, however, they are very often found in the same phanerogamic community.

On British marshes where the soil is rather sandy a *Vaucherietum* can be distinguished dominated by *V. sphaerospora,* but where the phanerogamic vegetation is very dense or heavily grazed by animals the algal vegetation is poor, e.g. south and west coast marshes. The Sandy Chlorophyceae and *Vaucheria thuretii* have a wide distribution, as also the *Catenella-Bostrychia* community, while the pan flora appears to be richest in East Anglia and Holland.

One of the more interesting features of the algal vegetation of salt marshes is the occurrence of the marsh fucoids. These are peculiar forms which are either free living on the marsh or else embedded in the mud, and they must all at one time have been derived from the normal attached form. Sometimes they bear a fairly close resemblance to the attached form but in other cases they have been very considerably modified, and it is only the existence of intermediate forms that enables us to indicate the normal type from which they came.

East Anglia is essentially the home of the marsh fucoids, although Strangford Lough in Ireland is also extremely rich. In Norfolk, considerable areas can be found occupied by *Pelvetia canaliculata* ecad *libera,* while the three marsh forms of *Fucus vesiculosus,* ecads *volubilis, caespi-tosus* and *muscoides* are also abundant, the last two being embedded in the soilf.

There are three other loose lying marsh forms derived from *Fucus vesi-culosus* confined to the Baltic, e.g. ecads *nanus, subecostatus* and *filiformis.* A small, crawling marsh form derived from *F. ceranoides* has been described from the Irish and Dovey marshes, and another larger, free living one from Strangford Lough in Ireland; like many others of this type it is profusely branched, fertile conceptacles are rare and, when present, are invariably female. *F. spiralis* vars. *nanus* and *lutarius* are other marsh derivatives, while *Pelvetia canaliculata* not only gives rise to ecad *liber a* but also to a small embedded form, ecad *radicans,* which has been recorded from the Dovey

There is another form, ecad *coralloides,* from Blakeney and Cum-brae marshes which requires further study of its status. *Ascophyllum nodosum* var. *minor* is a dwarf, embedded variety; ecad *mackaii* of the same species is a free living form found on American salt marshes, in Scotland and on the shores of Strangford Lough in Ireland, while ecad *scorpioides* is a partially embedded form found on the Essex marshes, on the shores of Strangford Lough and in eastern North America.

All these forms probably originated as a result of vegetative budding, although it is also possible that they have developed from fertilized oogonia that became attached to phanerogams on the marsh. There is definite evidence that *Ascophyllum nodosum* ecad *scorpioides* arises by vegetative budding from fragments of the normal plant, while it has been suggested that conditions of darkness or lowered salinity may be favourable for the development of ecad *mackii.*

In the southern hemisphere an interesting free living form of the southern fucoid *Hormosira banksii* has been recorded from mangrove swamps in New Zealand, though it does not form extensive communities.

As a group the marsh fucoids are characterized by:

1. Vegetative reproduction as the common means of perpetuation.
2. Absence of any definite attachment disc.
3. Dwarf habit.
4. Curling or spirality of the thallus.
5. In the species derived from *Pelvetia, Fucus vesiculosus* and *F. spiralis* that have been investigated it has been found that the three-sided juvenile condition of the apical cell is retained throughout life, and division in the megasporangia is at best only partial.

It is suggested that these features are due to:

a. Exposure, which results in a dwarfing of the thallus;
b. Lack of nutrient salts which induces a narrow thallus;
c. The procumbent habit and consequent contact with soil causes spirality because growth takes place more rapidly on the side touching the soil.
d. The cause of sterility may either be a result of the high humidity or, more probably, because of the persistence of the juvenile condition as represented by the apical cell and cryptostomata.

In New Zealand apart from the free living *Hormosira,* there is a *Gracilaria secundata* f. *pseudo-flagellifera* community which is widely distributed over mud flats and in mangrove swamps. It probably also occurs in Australia and would seem to be the only endemic southern hemisphere community so far recorded. The plants of *Gracilaria* are not originally free living but are attached to shells of molluscs buried in the mud.

The mangrove swamps represent an important habitat in the tropics and subtropics where they form the equivalent of the salt marshes. There are characteristic algal genera of mangrove swamps which are to be found throughout the whole pantropical belt, though the actual species may vary. Species of *Bostrychia* and *Catenella* form a turfy covering on the mangrove trunks and pneumatophores (aerating roots).

At a lower level on the same organs there is a mixture of *Caloglossa* spp, *Murrayella periclados, Poly-siphonia, Bryopsis* and *Centroceras clavulatum.* Numerous epiphytic species of *Enteromorpha, Lyngbya, Rhizoclonium* and diatoms occur on these turfy algae.

In the stream beds of the New Zealand mangroves, species of *Mono-stroma* can be found at certain seasons. A Cyanophycean community is also widespread on the pneumatophores of mangroves, and on the mud of New Zealand mangrove *Vaucheria* can form green patches. Similar green patches are formed by *Caulerpa verticillata* in tropical mangrove swamps.

In temperate regions beds of eel grass *(Zosterd)* are frequently associated with salt marshes. The tropical counterpart of submerged halophytic phanerogamic meadows dominated by *Halophila, Posidonia* and *Thalassia* are associated with coral lagoons and sheltered waters. Very few algae are specifically associated with such meadows though a variety of species, especially small brown and red algae are abundant. There are a few examples of specific relationships, e.g. *Smithora naiadum* occurring on *Phyllospadix* on the Pacific Coast from Washington to California.

The salt-marsh pan is the marsh equivalent of the rock pool. The number, shape and size of these on different salt marshes varies very considerably, but they generally contain a certain number of algae, especially in pans which occur on the lower marshes. Although some authors deny the existence of a true persistent pan flora Chapman and have shown that in one area at least a definite pan flora does persist, and that many of the species comprising it reproduce during the course of their existence.

On the lower marshes the pan flora is commonly composed of Chloro-phyceae, while with increasing marsh height the Chlorophyceae element decreases and the Cyanophycean element increases. A few of the constituent members, e.g. *Monostroma,* are seasonal in appearance, and on some marshes there are pans which contain algae that are normally associated with a rocky shore, e.g. *Colpomenia, Polysiphonia, Striaria.* These persist from year to year in spite of the stagnant conditions, and when compared with the habitats occupied by the same species on a rocky coast it is found that they are probably growing at an unusually high level.

A comparison of the Norfolk marsh flora with that of a comparable rocky coast leads to two generalizations which are probably valid for other marsh areas:

1. Some species that are littoral on a rocky coast are found growing at lower levels, usually sublittoral, on the marsh coast. This must be ascribed to the lack of a solid substrate at the normal higher levels.
2. Other littoral species of the rocky coast are found growing at higher levels on the marsh coast. Some of these occur in pans or in the streams where they are continually covered by water. In the case of a few species actually growing on the marshes it is probable that they are enabled to do so because the phanerogams provide protection from desiccation.

One feature of salt marsh areas are the lower mud flats. These can bear an embedded carpet of *Vaucheria* but in other places, particularly estuarine areas such as the Severn in England or the Avon in New Zealand, the muddy banks contain a population of *Euglena.* This can be so dense that during exposure in daylight the banks become a deep-green colour from the emergent organisms. Before the tide returns or at night the flagellates reburrow back into the mud.

Very little work has been done on the diatom flora of salt marshes. Round has demonstrated that two distinct groups exist, one associated with high marshes and the other with low marshes, and within these groups there are communities which are related to the different types of habitat available.

It has been suggested that light and space relations, rather than factors relating to level, influence the distribution of the various species. While this may have some truth there is no doubt that the increasing height of a marsh with its consequent greater exposure does nevertheless effectively determine the upper height to which many plants can go. The species to be found on the higher marshes in Norfolk are either fucoids or gelatinous Cyanophyceae, both of which have the power of retaining moisture.

The more delicate Chlorophyceae are more or less confined to the lower levels. On the other hand a dense phanerogamic vegetation, such as one finds on the south coast marshes where the tall *Spartina townsendil* must lower the light intensity considerably, does reduce the quantity of algal vegetation. A similar state of affairs has been observed on the grass covered marshes of New Zealand.

The distribution in space (e.g. among the different phanerogamic communities) and time of the marsh communities recorded from Norfolk. Some of the smaller communities, e.g. *Rivularia-Phaeococcus* and gelatinous Cyanophyceae, are apt to be overlooked in summer because the constituent species shrivel up so much or else because the colonies become covered by salt efflorescence. An examination of the distribution of the various communities on the Norfolk marshes showed that five communities are each confined to one type of habitat. This relationship may be due to:

1. Association with a particular phanerogamic

community, e.g. *Phormidiwn autumnale* (IX) with *Halimione portulacoides.*

2. Dependence on certain edaphic conditions, e.g. Muddy Chlorophyceae(Ic).
3. Dependence upon the physical character of the environment, for example slope, exposure, wave action, e.g. Marginal Cyanophyceae (III) Vernal *Vlothrix* (IV) and the Pan Association (XVI).

The factors operating on salt marshes are in many respects similar to those found on a rocky coast. The principal factor is the tidal one with all the derived factors that are associated with rise and fall of the tide (i.e. salinity changes, water loss, temperature changes, metabolic changes). Since salt marshes are restricted to protected coasts, there are not the changes that take place with transition to an exposed rocky coast.

10
Soil Algae and Symbiosis

SOIL ALGAE

The study of soil algae, as such, began seriously at the commencement of the nineteenth century with the works of Vaucher, Dillwyn, Agardh and Lyngbye. In 1895 Graebner, in a study of the heaths of Northern Germany, gave the first account of soil algae as ecological constituents, and subsequently many ecologists have shown that soil algae are pioneers on bare primary or secondary soil where they prepare the ground for the higher plants that follow. An agricultural, manured soil also has a rich flora, while the same species are to be found in unmanured soils, though not in such numbers.

This flora consists of about 20 species of diatoms, 24 of Cyanophyceae and about 20 species of Chlorophyceae. As might be expected, persistently damp soils have a more varied and extensive collection of algae than have generally dry soils. In the case of the diatom component of the soil flora it has been found that it is more abundant when the soil is rich in phosphates and nitrates, so that the soil salts may also be of importance.

It appears that the method of studying soil algae is of great importance. In the early years dilution techniques were employed as was direct examination. There seems little doubt that many species can be missed by this technique. Various culture methods have since been evolved, using soil samples on sand and mineral solution, on agar and mineral solution or a moist-plate culture of the soil itself. Both Round and Forest *(loc. cit.)* conclude that the moist-plate culture is the most satisfactory.

Whilst the distribution of many of the common species of soil algae is almost world wide, nevertheless some species appear typical of certain soil types. Thus peat soils are characterized by an abundance of desmid species, Cyanophyceae such as *Chroococcus* and *Gloeocapsa*, and diatoms.

With a rich alkaline loam these algae are replaced by others, especially coccoid algae, Ulotrichales (*Hormidium, Ulothrix*), filamentous Cyanophyceae (*Oscillatoria, Phormidiurri*) and diatom genera different from those on peat soils. On the other hand, soils of one kind throughout the world may possess a similar flora, e.g. arid soils from Saudi Arabia, Mexico, Arizona, Utah and Israel all have much the same species: *Chlorosarcinopsis* was found in nearly every sample and *Chlorococcum diplobionticoideum* was almost as frequent.

In temperate and monsoonal regions the soil flora may vary with the season or monsoon. Soils, however, are not permanent and are continually in a process of maturing. New soil is rapidly colonized by algae, but as colonization and succession of higher plants takes place, the amount of organic matter increases, changes occur in the nutrient status and the composition of the algal flora changes accordingly. There seems little doubt that the soil flora has an important biological significance, but at present much more experimental data is required. It may well be that the biological character of a soil will prove to be of more importance than its chemical or physical character.

SUBTERRANEAN ALGAE

There are great fluctuations in the numbers of the different species that compose the flora, but there are no species in the lower layers of the soil which do not also occur in the surface layers. The most actively growing algae are probably confined to the top inch or so of soil, where some of the species are motile, e.g. *Euglena, Chlamydomonas, Oscillatoria, Phormidium* and pennate diatoms such as *Na'vicula, Pinnularia, Stauroneis.* On the surface of damp soils in rain forest climates macroscopic, mucilaginous colonies of *Nostoc* can be found. The most advanced soil alga, morphologically, is *Fritschiella.*

In the soil there are two main groups of algae: first there are ephemerals which only appear in quantity when conditions are favourable, e.g. *Chlamy-domonas, Euglena.* Secondly there are the perennials, mainly diatoms and Cyanophyceae.

Within the soil community Protozoa and rotifers may devour algae and their grazing may be selective. There are also soil fungi, especially members of the Chytridiaceae, that attack algae, but a study of both these interrelationships has scarcely begun.

Lund studied 66 different soils in Great Britain and found that, except on acid soils, the Chlorophycean component was the largest, but it was followed closely by the diatoms with the Cyanophycean element third. In the case of cultivated soils, however, the Cyanophycean element became much more important. Very little work has so far been carried out on the soils of tropical regions. A study of the soils of savanna, steppe and maxanga in the Congo yielded a flora almost entirely composed of Cyanophyceae, though in true savanna algae were rare, probably because of competition for space and light.

The growth of soil algae has been a source of interest and experiment for a number of years. In 1926 Roach found that *Scenedesmus costulatus* var. *chlorelloides* would grow in the dark when provided with mineral salts and glucose. A more recent contribution has demonstrated that at least a third of the soil algae tested were capable of fair, good or excellent heterotrophic growth in darkness in the presence of one or more carbon sources. The successful algae included species of *Neochloris, Spongiochloris, Chlorosarcinopsis* and *Friedmannia israeliensis.*

The maximum depth to which algae have been recorded in soils is 2 m. In Greenland soil algae have been found down to a depth of 400 mm and, because burrowing animals are absent, their presence there can only satisfactorily be explained by the action of water trickling down cracks. In the lower layers of the soil the algae exist in a dormant condition, either as spores or as filament fragments.

The occurrence of algae in the lower soil layers depends on:

(1) Cultivation,
(2) Animals,
(3) Water seepag,
(4) Self-motility.

Mechanical resistance and lack of light are said to prevent

Cyanophyceae from moving down through their own locomotion, but even so some algae such as *Euglena*, may be able to move down by their own mobility. The effect of water seepage will depend on the heaviness of the rainfall, state of the soil, i.e. whether cracked or waterlogged, and the nature of the algae, i.e. whether or not they possess a mucous sheath that will adhere to soil.

With filamentous algae movement is helped by fragmentation or by the formation of zoospores. Many green algae are known to form zoospores when put in water after a period of drought, and hence one may presume that rain after a dry period will induce zoospore formation. Petersen has demonstrated that rain can carry algae down efficiently to depth of 200 mm, but the process is facilitated by the presence of earthworms, although these animals probably only operate indirectly in that they loosen the earth. Soil cultivation is obviously very important here.

Many soil algae, especially Cyanophyceae, can resist long spells of drought. When remoistened, bacteria usually first develop, then coccoid green algae and finally Cyanophyceae that quickly become dominant. *Nostoc muscorum* and *Nodularia harveyana* appeared from soil that had been dried for 79 years, and *Nostoc passerinianum* and *Anabaena oscillarioides* var. *terrestris* were obtained from a soil dried for 59 years.

In a study of the moisture relations of some soil algae Fritsch and Haines showed that:

1. There is a complete absence or paucity or large vacuoles.
2. In an open dry atmosphere nearly all the sap is retained.
3. When the filaments dry up, contraction of the cell is such that the cell wall either remains completely investing the protoplast or else in partial contact with it, thus ensuring that all the moisture which is imbibed will reach the protoplast.
4. During a drought there is, as time goes on, a decreasing tendency for the cells to plasmolyse and there are also changes in the permeability of the cell wall, while the excess of moisture normally brings about changes in the reverse direction. The majority

of cells which plasmolyse lack the characteristic granules, mainly of fat, that are to be found in most terrestrial algae.

5. Those cells which survive after drought do not possess any vacuoles, and have instead a rigid, highly viscous protoplasm that is in the gel condition.
6. If desiccation is rapid most of the cells will die but some will plas- molyse and retain their vitality in that state for weeks or months. Thus no species is likely to disappear from a flora during the rapid onset of drought.
7. If desiccation continues the number of living resting cells will remain constant for several years.

Apart from moisture, temperature in extreme climates may be important.

Diatoms can survive very low temperatures, - 80° C for 8 days or - 192° C for 13 hours, while dry spores of *Nostoc* sp. and *Oscillatoria brevis* can survive -80° C, though if they are moist temperatures below - 16° C are fatal. So far as the algae of tropical soils or thermal regions are concerned the dry spores of *Nostoc* sp. and *O. brevis* can tolerate 2 mins. at 100° C., the wet spores 20 mins. at 60-70° C, and the vegetative filaments 10 mins. at 40° C, this latter being a temperature that is frequently reached on open soil surfaces in the tropics. In general it seems that the distribution of soil algae is related to the physico-chemical characters of the soil, texture, water capacity etc.

Soil algae also add certain compounds to the soil, either through death or by secretion from the cells. Cyanophyceae and Chlorophyceae have been shown to release polysaccharides in culture and this probably occurs in nature. Those Cyanophyceae that fix nitrogen play an important part. Certain algae also produce growth substances and antibiotics, e.g. *Chlorella*, which may affect the soil flora. It is these compounds that must play some part in the success of the soil culture (Erdschreiber) media for many algae.

SYMBIOSIS

The most striking and well known examples of symbiosis involving algae are provided in those cases where the plants are associated with animals, especially Coelenterates, or with fungi, as in the Lichens. Animal symbiotic associations are known from the Protozoa through to the Prochordates (Tunicata). Apart from these examples, however, there are other cases which are not so well known, mainly because they are not so common. *Gloeochaete,* is a colourless genus of the Tetrasporaceae, the members of which possess blue-green bodies that look like chromatophores, though they are really a symbiotic member of the Cyanophyta.

Glaucocystis is a colourless genus of the Chlorococcales in which a symbiotic member of the Cyanophyta also forms the blue-green 'chromatophores' that appear as a number of curved bands grouped in a radiating manner around the nucleus. In this case the illusion is further enhanced because the endophytes break up into short rods at cell division. It has so far proved impossible to grow the blue-green alga separately and it may thus have lost its power of independent growth.

In the flagellate *Cyanophora paradoxa* the algal symbiont also belongs to the blue-green algae, but it lacks the usual double-layered wall and has a thin single membrane instead. The cyanelle has been named *Cyanocyta korschi-koffiana. Geosiphon,* which is variously regarded as a siphonaceous alga or as a Phycomycete, possesses small colonies of *Nostoc* enclosed in the colourless, pear-shaped vesicles that arise from an underground weft of rhizoidal threads. Reproduction by the formation of new vesicles is said to occur only in the presence of the *Nostoc.* The presence of chitinous material in the vesicular wall suggests a fungal nature for *Geosiphon,* the vesicles perhaps being galls that are formed on the threads as a result of the presence of the alga.

The principal genera taking part in lichen synthesis are *Nostoc, Scytonema, Gloeocapsa, Cephaleuros* and *Trentepohlia.* Under normal conditions the partnership is truly symbiotic, but under abnormal conditions the fungus may become a parasite and devour the algal component. The mycobiont cannot normally live apart from the alga, but most of the phycobionts appear capable of an independent existence.

The green bodies which are found associated with the cells of Coelentrates and Radiolarians are usually placed in 'form' genera, *Zoochlorella* and *Zooxanthella.* Most of the species are now known to be dinoflagellates, e.g. *Symbiodiniwn microadriaticum* in the coral *Cladocora* and the giant Pacific clams of the genus *Tridachna*. In the freshwater sponges *Chlorella parasitica* imparts the characteristic green colour, though *Gongrosira* and *Aphanocapsa* also occur in freshwater sponges.

The non-motile cells of the coral dinoflagellates are usually found in the peripheral layers of the polyps, the host larval stages usually lacking the alga. It is the existence of the independent motile phase that has enabled the algae to be taxonomically determined. The function and relations of these symbiotic algae have been discussed over the years, and on the whole there appears to be evidence for the alga obtaining food, especially nitrogenous compounds, from the animal, and the animal oxygen and carbohydrates from the alga.

In the corals the algae are regarded as speeding up the rate of removal of waste products and this involves the deposition of waste calcium carbonate. The problem of the relationship between algae and animals is by no means completely worked out, and it is not impossible that in some cases the animal is really a parasite of the alga. Some colourless Cyanophyta *(Anabaeniolum, Oscillospird)* typically occur in the digestive tract of mammals (including man), where they live as saprophytes.

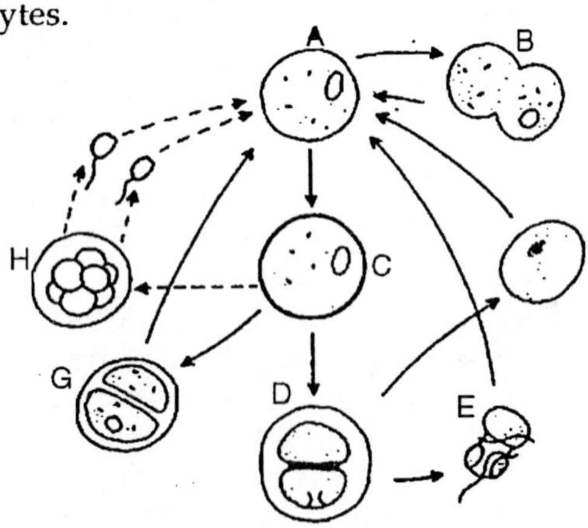

Fig. Life cycle of *Symbiodinium adriaticum.*

When cultured outside in the light they become autotrophic. In the worm-like *Convoluta roscoffensis* with its algal associate *Platymonas,* the animal cannot live unless infected with the alga though under certain conditions it also digests the algal cells. Exactly how the algae re-infect the host animal or are transmitted is not well known, though there are clearly various potential methods.

[A, vegetative cell; B, binary fission of vegetative cell; C, vegetative cyst; D, zoosporangium with zoospore; E, zoospore; F, aplanospore; G, cyst with two autospores; H, cyst with developing gametics; I, isogametics.]

Examples of a looser form of symbiosis (almost a commensalism) are provided by *Anabaena cycadearum* living in the cortex of *Cycas* roots, and *Anabaena azollae* found in the leaves of the water fern, *Azolla filiculoides.* Species of *Nostoc* found in the thalli of the liverworts *Blasia* and *Anthoceros* are probably no more than space parasites obtaining shelter. This is not so, however, with *Nostoc pruniforme* that lives in the petioles and stems of large-leaved tropical *Gunnera* species, because the host probably gains from nitrogen fixation by the alga.

Epiphytism is extremely common among the algae, whilst there are also a number of epizoic forms. One may also find endophytic species, e.g. *Schmitziella endophloea* in *Cladophora pellucida,* species of *Endoderma* in various algae, and endozoic species such as *Rhodochorton endozoicum* in hydroid sheaths. *Blastodinium* is an endophyte of Crustacea and worms.

The origin of the symbiotic habit among the algae is probably to be explained as cases of epiphytism or parasitism, and in the former the relationship between host and epiphytes became more intimate while in the latter case the host evolved mechanisms that enabled it to tolerate the parasite and take advantage of it.

Many examples of algal parasites are known. Among the Chlorophyta there is *PhyUosiphon arisari, Chlorochytrium lemnae* inhabiting the fronds of *Lemna trisulca* and species of *Cephaleuros. Notheia anomala* is at least a hemi-parasite if not a holo-parasite in the Phaeophyta, while in the Rhodophyta some parasitic forms, e.g. *Harveyellamirabilis Janczewskia* spp., *Faucheocolax, Colacolepis, Hypneocolax,* have all become highly modified as a result of their parasitic habit. Although these parasites have been known for a long time no intensive work has been carried out on the nature of the parasitism.

11
Algal Physiology

It is impossible to separate algal physiology completely from some aspects of ecology, and many of the problems of algal physiology are those of plant physiology. It is essential to distinguish between physiological phenomena primarily found in algae (e.g. certain pigmentations, phototaxis, luminescence, hydrogen adaptation) and general plant physiology in which algae are experimental organisms. It is intended to give only a very brief outline, as algal physiology has been comprehensively discussed by Lewin and specific topics are annually discussed in the Annual Reviews of Plant Physiology and Microbiology, and Advances in the Microbiology of the Sea.

Apart from physiologico-ecological studies our knowledge of algal physiology stems largely from the study of bacteria-free algal cultures. Because unicellular organisms are much easier to handle than multicellular algae, the net result is that much of our knowledge is at present centred around these forms. The use of these forms as a means of extra food in arid and nutritionally poor areas may be important in the future and micro-algal cultures may also be significant in space vehicles. Some of the larger green algae, *Enteromorpha, Ulva, Caulerpa, Halimeda, Valonia, Acetabularia,* a few brown algae, e.g. *Pylaiella, Ectocarpus* and a few red algae, e.g. *Porphyra, Traillatella, Callithamnion* have recently been grown and maintained successfully in culture.

PHOTOSYNTHESIS

Some mention has already been made of the considerable variation in pigment composition to be found in the different algal classes. In the last few years considerable work has been carried out on these pigments in order to determine the part they play in the absorption of light for photosynthesis. In the larger marine algae this is important in the depth distribution of the green, brown and red algae.

Fucoxanthih in the Phaeophyta, diatoms and Chrysophyta transfers absorbed light energy to chlorophyll with very high efficiency while peridinin in the Pyrrhophyta is also an efficient agent of energy transfer. Other carotenoids however have a low efficiency of energy transfer and 8-carotene now appears to play an important role in protection against photosensitization. The phycocyanin and phycoerythrin biliproteins also show a very high efficiency of energy transfer in photosynthesis.

When light passes through water it is differentially absorbed and it seems that the red-orange and violet parts of the spectrum are more easily absorbed. In the sub-littoral depths, the green and green-orange portions are the principal penetrating light wavelengths. It is in this region that red and brown algae predominate while the Chlorophyceae, with maximum absorption in the blue and red-orange predominate near the surface.

Thus pigmentation patterns do have definite ecological importance. The presence of biliproteins, fucoxanthin and peridinin is primarily genetically controlled, and these pigments are rarely, if ever, completely absent. It must be emphasised, however, that the pigment ratios are greatly affected by light intensity, light quality and nutrient availability.

Furthermore zonation is profoundly influenced by the general problem of exposure and submergence, e.g. many brown algae are intertidal. Engelmann's early views of chromatic adaptation are at least partially borne out. The phrase 'chromatic adaptation' should however be used with care and indeed "chromatic control' might be a better term, restricted to such phenomena as the biliprotein variations in *Tolypothrix*.

There is no evidence that the mechanism of photosynthesis in the green algae differs materially from the mechanism in higher green plants. From the work that has been carried out it seems that the rate of photosynthesis in marine algae, such as fucoids, is lower than that of land plants. It is also evident that in the case of larger littoral algae the capacity to photosynthesise is greatly reduced during periods of exposure.

Under normal conditions water acts as the electron donor for photosynthesis but certain algae are capable of photoreduction using hydrogen after a period of adaptation under anaerobic conditions. The marine algae require a long period of adaptation. The phenomenon is dependent upon the presence of a latent hydrogenase which is activated under dark anaerobic conditions.

There is also evidence that some algae, like certain bacteria, can utilize hydrogen sulphide as the hydrogen donor in place of water. *Oscillatoria, Synechococcus, Pinnularia* (diatom), and *Scenedesmus* can all utilize H_2S, though *Synechococcus* and *Scenedesmus* use molecular hydrogen preferentially.

Various factors, especially light and temperature affect photosynthesis. With increasing light intensity there is light saturation and oxygen output then levels off. With shade algae, such as *Pterocladia capillacea* light saturation can be at quite low values, though even this point can be modified by temperature and CO_2 concentration.

When light is limiting, temperature exerts little effect, but once light is no longer limiting then increasing the temperature may raise the rate of photosynthesis at the level of light saturation. In the case of *Cystoseira barbata,* summer photosynthesis with maximum water temperature and light intensity of 0-16-0-20 cals cm ^{2} min^{-1} is around 80-135 mg 0_2/ 100 g fresh alga in one day. From October to January, with reduced light intensity and low temperatures photosynthesis is much lower.

Work in New Zealand has shown comparable phenomena with local species. Certain species of *Ulva, Chaetomorpha* and *Enteromorpha* that occur in the littoral can tolerate changes from 14° to 32° C and back to 14° C with little or no reduction in the rate of photosynthesis.

RESPIRATION

Detailed studies of the respiratory process have shown that it follows the Emden-Meyerhof-Parnas glycolysis path and this is followed by the Krebs cycle, in *Chlorella, Scenedesmus, Ulva, Myelophycus* (Phaeophyceae) and *Gelidium*

(Rhodophyceae). In the presence of light, respiration appears to follow a different path in *Nitella, Iridaea, Polyneura* and *Sargasswn* and this aspect wants further study.

Respiration does not appear to follow the normal path in *Cylindrospermwn* (Cyanophyceae) nor does this alga appear to have the enzymes associated with the Krebs cycle. The respiration rate of some algae, especially fine and delicate species, is increased by lowered salinity and decreased by salinities in excess of sea water. Anaerobic respiration (fermentation) has been reported for a number of algae, e.g. *Scenedesmus,* the colourless *Prototheca zopfii.*

CHEMOTROPHY

Chemotrophy involves the utilization of -inorganic (-lithotrophy) or organic (-organotrophy) compounds as a source of energy. Certain algae appear to be obligate chemotrophs, e.g. *Gonium quadratum, Stephanosphaera pluvialis,* require acetate in the light for growth.

Many are facultative chemotrophs. In the dark they utilize sugars (e.g. glucose) organic acids (e.g. acetate) and occasionally alcohols and amino acids. Facultative chemotrophy (or dark heterotrophic growth) is established for members of the Chlorophyceae (e.g. *Volvocales),* Xanthophyceae *(Tribonema),* Bacillariophyta (e.g. *Navicula pelliculosa),* Chrysophyta *(Ochro-monas, Prymnesium),* Euglenophyceae *(Euglend),* Cyanophyceae *(Tolypothrix).* Colourless algae such as *Prototheca* (colourless *Chlorelld), Chilomonas* (colourless Cryptophyceae), *Polytoma* (colourless Chlamydomonadaceae) are of course obligate chemotrophs.

The green flagellate *Chlamydobotrys* is interesting because it can grow in red light when supplied with acetate, but not in green or blue light. This is an example of photochemo-organotrophy. *Ochromonas malhamensis* grows very slowly in the light, because of a very low rate of photosynthesis and an additional carbon source (e.g. glucose) is required for rapid growth.

At the other end of the scale is obligate phototrophy, which means the complete inability to grow in the dark. This phenomenon is manifested by certain *Chlamydomonas* spp;

Monodus (Xanthophyceae), certain Cyanophyceae and numerous Bacillariophyceae and Dinophyceae.

Oxidative assimilation, in which a portion of the organic substrate (e.g. acetate) is synthesized into cell material, and part used as an energy source in respiration, has been demonstrated in the colourless *Prototheca,* and also in *Chlorella pyrenoidosa, C. vulgaris* and *Scenedesmus quadricauda,* and it is probably even more widespread.

Facultative chemotrophy may be important in the metabolism of deep growing marine algae and in situations where the available light may be negligible during long periods of time.

OTHER NUTRIENTS

Phosphorus is one of the essential elements for the algae but the amount necessary can vary very widely from less than 0.005 mg $P/10^{3}\ m^2$ for *Dinobryon* to 0-45 mg $P/10^{3}\ m^2$ for *Asterionella formosa.* In the presence of phosphate deficiency there can be an accumulation of fat. Certain algae, e.g. *Euglena, Volvox, Pandorina, Spirogyra* are capable of accumulating phosphates.

Sulphur is probably assimilated as sulphate though under strong reducing conditions it may be present as hydrogen sulphide. There appears to be some connection between the presence of sulphur and the uptake of silica by diatoms. The uptake of sulphate by *Fucus vesiculosus* is apparently related to light intensity and in this alga it is incorporated into fucoidin and in other algae it appears generally in various organic compounds.

Silica is essential for diatoms and certain members of the Chrysophyta and Xanthophyta. It is normally present in natural waters in adequate supply but it can be limiting at certain times and this affects the development of some planktonic species. At least 0-5 $mg/10^{-3}\ m^2$ are required for successful growth of *Asterionella formosa* and 25$mg/10^{-3}\ m^2$ for optimal growth of*Fragilaria crotonensis.*

Calcium and magnesium are normally both present in sufficient quantity. Certain algae, e.g. Calcareous Siphonales, Dasycladales, Dictyotales and Corallinaceae are capable of accumulating calcium and depositing it in the form of a calcareous skeleton.

Sodium is not normally a requirement for the growth of algae though it has been shown to be essential for certain Cyanophyceae, e.g. *Anabaena variabilis, Anacystis nidulans* and *Nostoc muscorum*. In the case of other algae excess sodium can be inhibitory and a sodium efflux pump has been demonstrated for *Hormosira banksii, Ulvalactuca, Valonia macrophysa, Halicystis ovalis* and *Acetabularia*. Potassium is an essential element for all algae and in marine algae it is accumulated by an influx pump against a gradient, e.g. *Ulva*. Under low concentrations of potassium both growth and photosynthesis are low and respiration is high.

Iron and manganese are essential elements because of the part they play in metabolic processes. Seasonal variations in manganese can be reflected in the composition of the phytoplankton. In the same manner that higher plants require certain trace elements so culture work has shown that these are essential for the successful growth of algae.

Some of the trace elements concerned are molybdenum, copper, boron and cobalt. The red seaweed *Asparagopsis armata* grows very much better with both iodine and arsenic, and boron is important for Cyanophyta. Members of the Phaeophyceae and Rhodophyceae have the capacity to accumulate iodine in considerable quantities.

VITAMINS

Culture work has shown that three vitamins, cobalamin (B_{12}), thiamine and biotin are needed alone or in combination by many algae. Species requiring external growth factors, e.g. vitamins, are called auxotrophs. Many algal species, distributed among nearly all algal divisions other than the Phaeophyta are known to require an external source of vitamin B_{12} and at least 40 species are known to require thiamine or a part of the molecule. Biotin is required by only a few algae, all of which require one or both of the other two.

Growth rates of many auxotrophic algae are very sensitive to low concentrations and such algae (e.g. *Euglena gracilis, Ochromonas, Monochrysis)* have been used in the bioassay of B_{12}. In general it may be said that algae with strong saprophytic and holozoic tendencies require vitamins.

In some of the higher algae, e.g. *Laminaria hyperborea, Ascophyllum,* the growth substance niacin varies seasonally, reaching up to 35 mg/g dry wt. in spring and falling to 15 mg/g dry wt. in autumn. Some species of algae contain as much vitamin C as do lemons, though the content is affected by season, temperature, location and pH. Vitamins D and E also appear to be present in at least some algae.

There is some evidence that the growth of marine algae may be dependent on organic substances present in sea water, though the nature of these substances is not fully understood. Surface sea water, has substances necessary for the germination of sporelings of *Ulva* and *Entero-morpha.*

Fucoids in the northern hemisphere are known to excrete phenolic compounds and so do *Hormisira* and *Ecklonia* in the South Pacific. Other organic substances are excreted by *Rhodymenia palmata, Spongomorpha arcta* and *Halosaccion ramentaceum.* Not all marine algae excrete because there is no evidence of any compounds from *Pterocladia capillacea.*

An interesting phenomenon associated with some of the larger marine algae and the significance of which is not yet determined, is the production of carbon monoxide. In certain algae *(Nereocystis, Pelagophycus)* the carbon monoxide accumulates in the pneumatocyst. Algae in which carbon monoxide production has been demonstrated in fresh material include *Egregia menziesii,* and *Macrocystis,* it is also recorded for chopped material of *Laminaria Calliarthron, Rhodomela larix* and *Iridophycus flaccidum.*

NITROGEN METABOLISM

Algae utilize nitrate, ammonium salts and organic nitrogen as the main sources for nitrogen although ammonium ion appears to be preferentially absorbed over nitrate. Some algae can use nitrites and others are capable of fixing nitrogen from the atmosphere. Certain marine algae have the capacity to accumulate nitrates against a gradient; thus *Valonia* accumulates nitrates up to 2000 times the concentration in sea water and *Halicystis* up to 500 times.

Detailed studies of nitrogen absorption by various algae have led to the conclusion that irrespective of source the nitrogen is eventually incorporated into the metabolic pathways as ammonia. It appears that the amination of a-keto-glutaric acid, yielding glutamic acid, represents the major pathway of ammonia incorporation into amino acids with a lesser one via aspartic acid. The actual sites of protein synthesis vary in different species and it is not necessarily associated with the nucleus; thus enucleated *Acetabularia* continues to synthesise protein in both cytoplasm and plastids for at least three weeks.

The establishment of bacteria-free cultures has enabled workers to show that some twenty or more species of Cyanophyceae are capable of 'fixing' atmospheric nitrogen. This includes all genera in families that form heterocysts, e.g. *Nostoc, Anabaena, Cylindrospermum, Calothrix, Tolypothrix, Hapalosiphon, Anabaeniopsis* and *Mastigocladus.* Some species can fix nitrogen in the dark if supplied with sugars but others require light for the process. Fixation only takes place when there is little, if any, source of readily available combined nitrogen, e.g. nitrate, ammonium salts.

The process also apparently requires the presence of traces of molybdenum and is also more efficient in the presence of bacteria. In marine environments the phenomenon in *Calothrix scopulorum* and *Nostoc entophytum* is not affected by variations in salinity. The amount 'fixed' is highest in winter and lowest in summer.

The fixation of free nitrogen by Cyanophyceae is of considerable importance in the culture of rice and continual crops are due to fixation by *Anabaena* and *Cylindrospermwn gorahporense.* Watanabe has shown that *Tolypothrix tennis* can fix nitrogen at a rate of 0-248 mg/m^2 in 12 h which could add as much as 2,900 lb of nitrogen per acre in a year. This alga is now grown in culture and is used to seed the rice fields; in India, innoculation by Cyanophyceae is an effective substitute for fertilizing with ammonium sulphate.

GROWTH HORMONES

Auxin (indole-acetic acid) has been reported in *Valonia macrophysa.*

Laminaria agardhii and in a number of Pacific coast algae, including *Macro-cystis, Desmarestia* and *Fucus evanescent.* I.A.A. promotes growth in *Codium decorticatum, Ulothrix subtilissima* and *Acetabularia mediterranea.* When associated with kinetin sporeling growth and initiation of normal filaments from green island cells in white thalli of *Ulva lactuca* was much better than kinetin by itself which only induced rhizoid formation. Gibberellic acid alone merely induces increased elongation of the thallus, though it promotes a 7-fold growth increase in *Ulothrix.* Not all algae appear to require growth hormones, e.g. *Prasiola stipitata.*

Growth substances have been detected in the eggs, sperm, and fruiting tips of European species of *Fucus,* and the presence of these may be partly responsible for the fact that the first rhizoid of a germinating zygote grows out on a side adjacent to neighbouring eggs. This is not the only factor as the appearance of the rhizoid is also affected by pH, and by unilateral illumination; furthermore blue rather than red light orientates the cleavage wall. When ova of *Fucus* or *Cystoseira* are passed through capillary tubing rhizoids are formed at a pole which suggests that pressure can play some part as well.

In addition to growth-promoting substances there may be growth-inhibiting compounds. Such substances are probably produced by *Nostoc punctiforme, Nitzschia palea* and *Chlorella vulgaris.* The inhibitor from *Chlorella* is known as 'Chlorellin' and appears to be a mixture of fatty acids. It is active against bacteria and this is one reason why *Chlorella* is effective in purifying sewage effluent.

SEX HORMONES

Since 1926 it has been known that the gametes or filaments of certain algae secrete sexual substances into the surrounding water. This has been demonstrated for *Dasycladus clavaeformis, Tetraspora lubrica, Sphaeroplea, Vaucheria, Oedogonium, Chlamydomonas eugametos, Volvox, Protosiphon botryoides, Stephanosphaera pluvialis, Ectocarpus* and *Botrydium granulatum.*

CHEMICAL COMPOSITION

As may be expected, the cell-wall material and the major storage products are composed mainly of polysaccharides.

Apart from the Procaryota the principal wall polysaccharides are pectin (small amounts) uronic acid compounds, glucans, xylans, mannans and galactans, many being of commercial importance. In the larger Phaeophyceae alginic acid is present in the walls and in cells of certain Rhodophyceae can be found agar and carrageen which again are of commercial importance.

Starch, is the principal storage product of the Chlorophyta, Cryptophyceae, some Dinophyceae and Rhodophyceae (Floridean starch). The storage product of the Cyanophyceae is similar to the starches, but may be more closely related to glycogen. Storage reserves with principally glucoside linkages are found in the Phaeophyta (laminarin) Chrysophyta and Bacillario-phyta (chrysose or chrysolaminarin) and Euglenophyceae (paramylum). Other common reserve carbohydrates are trehalose, floridoside and manno-glycerates which are found in the Rhodophyceae, while polyhydric alcohols are common in the Phaeophyceae and Rhodophyceae. Dukitol is present in *Iridaea* and *Bostrychia* and mannitol is present in both groups. The presence of sorbitol *(Bostrychid)* and volemitol *(Pelvetia* and *Porphyrd)* has also been established.

Because of the commercial uses of some of these storage compounds a great deal of work has been carried out on the detailed chemical composition of algae, especially the littoral rock weeds (Fucales) and sub-littoral oarweeds (Laminariales) of Europe. Regular analyses of plant populations have been made so that the results represent changes in the average composition rather than those in a single individual. During the period of rapid growth (spring), fresh weight, ash, protein content and alginic acid increase while laminarin and mannitol behave in a reverse fashion.

The summer decrease in growth is correlated with the decrease NO_3 in the sea at that time. During this period, however, the high photosynthetic rate permits the accumulation of mannitol and laminarin. There is also a tendency to more than one maximum in mannitol and laminarin, due largely to the interaction of exposure, temperature and formation of reproductive organs.

The low laminarin content in July-August of Scottish plants is almost certainly the result of a lowering of photosynthetic activity brought about by increased desiccation. The degree of the variations depends upon habitat, especially for *Laminaria digitata* and *L. saccharina,* being least in the open sea and greatest in locks or enclosed bodies of water.

In *Laminaria hyperborea* the seasonal variations are almost entirely confined to the frond. With *L. saccharina,* gradients in mannitol and laminarin occur along the frond, the amounts appearing to be related to the age of the tissue. Mannitol and laminarin contents also vary with depth. In the case of littoral fucoids seasonal variations tend to increase with increasing frequency of immersion.

Outside Europe rather similar results have been reported for *Fucus vesiculosus, F. evanescens* and *Ascophyllum* from the Canadian Maritime provinces, except that laminarin tends to remain rather constant.

SAP, PERMEABILITY AND OSMOTIC PRESSURE

The Siphonales *(sensu latd)* and Charales have formed excellent material for studies in permeability, sap composition and osmotic pressure because of the size of their cells. There is considerable variation in the rate of penetration by the different ions; iodine, enters far more rapidly than chlorine, and potassium more rapidly than sodium but less rapidly than ammonia. The behaviour of salts and gases in the case of algae with a higher and more complex organization of tissues may well be different from the Siphonales *(sensu latd)* and Charales, but so far it has been but little studied.

It is known that there are influx and efflux pumps operating for sodium and potassium. Dilution of sea water brings about the escape of chloride and potassium ions from *Chaetomorpha linum* but the sodium ion is little affected. The entry and exit of ions is related to the availability of free space, which in the case of *Ulva lactuca* and *Cladophora gracilis* may amount to 20 per cent of the wall volume. Very little work has been done on free space in algal cell walls and this field demands further investigation. The behaviour of the sulphate ion is also peculiar because it appears to be almost wholly excluded by both *Valonia* and *Halicystis* but not by *Ulva* and *Entero-morpha.*

The cell sap of many freshwater algae has an osmotic pressure of about 5 atmospheres, whereas that of some marine algae, e.g. *Halicystis, Valonia, Nereocystis,* is only slightly higher than that of sea water. The osmotic pressure in Fucales would appear to be rather higher as is that of some of the Rhodophyceae. In general, however, the osmotic pressure in red algae is somewhat lower than that in green or brown algae. The tolerance of algae to changes in osmotic pressure (caused by water loss on exposure) may be important in the case of littoral seaweeds; some algae appear to respond almost instantaneously, e.g.. *Monostroma, Dunaliella salina.*

This suggests that cells of such species may be very permeable to the salts rather than impermeable to water. Changes in osmotic pressure affect respiration in *Laminaria, Hormosira banksii, Fucus serratus, Chaetomorpha linum* and other algae, but not in *Ulva, Porphyra* or *Fucus vesiculosus.* Photosynthesis is affected in *Fucus, Ulva* and *Hormosira,* being more than doubled in the first two genera when the sea water is diluted a third.

BIO-ELECTRIC PHENOMENA

Using large cells of the Charales, *Halicystis, Valonia* or *Hydrodictyon* it has been found that potential differences exist between the interior and the exterior of the cell, the latter usually being positive, e.g. *Chara, Nitella, Halicystis, Bryopsis, Hydrodictyon,* though in *Valonia, Ernodesmis* and *Chamaedoris* it is negative. It seems likely that the gradient of potassium ions is responsible for this potential in *Nitella* and *Hydrodictyon.* In other algae it seems probable that the potential is a result of an internal concentration gradient related to all the compounds within the protoplasm itself.

POLARITY AND MORPHOGENESIS

Polarity is a problem of considerable interest, especially in the case of coenocytic algae. Reference has already been made to determination of *Fucus* egg polarity, and spores of other algae, e.g. *Griffithsia,* are also polarized because rhizoid and shoot appear at opposite poles. Rhizoid formation can be induced by subjection to an electric current when rhizoids develop on the anode side of the thallus. Polarity also exists in *Enteromorpha* in relation to its powers of regeneration.

Thus when apical sections are cut from the thallus papillae appear on the morphological upper surface and rhizoids and papillae on the morphological loweir surface. It has been suggested that a gradient of substances is set up early in development and that it is not readily disturbed.

In *Bryopsis* accumulation of auxin in the lower half of the thallus brings about rhizoid formation. If plants of *Bryopsis* are cut, inverted, and the cut end treated with I.A.A. rhizoid formation takes place at the lower tips where LA.A. accumulates.

The formation of the cap in *Acetabularia* is also an example of polarity. Enucleate apical segments have a greater regenerating power than enucleate basal segments. In this alga there is now clear evidence of the production of morphogenetic substances by the nucleus in the rhizoid.

The genus *Acetabularia,* has proved an admirable genus for experimental work on morphogenesis. Important summaries of the great volume of work can be found in Hammerling, Keck, Brachet and Puiseux-Dao. Because of the large single nucleus in the rhizoid a study can be made of the influence of the nucleus on structure by its removal. It was found in 1931 that enucleate portions *of Acetabularia* could survive and regenerate and possessed a morphogenetic capacity related to the orginal nucleus.

When the nucleus is removed by cutting off the rhizoidal base, the stalk portion will grow a cap, or with *A. crenulata* more than one cap. If the stalk is cut into several portions normally only the apical one will elongate and it may form a cap at one or both ends. It is argued that the nucleus can produce a morphogenetic substance or substances which accumulate in the cytoplasm and exert their control even after the nucleus is removed. The greater regenerating power of the apical segments of the stalk suggests that there is a gradient of these substances.

Quite early on it was found possible to make grafts by inserting the cut stalk of one plant into the cut base of another. As a result grafts between species could be made and it also became possible to obtain bi-, tri- and quadri-nucleate plants.

Later it has been found possible to transfer single nuclei so that in the final plant there is no mixed cytoplasm.

Using these grafts it has been shown that the form of the cap is determined by the nucleus present, and if more than two nuclei were present the cap morphology tended to be characteristic of the preponderant species nuclei, e.g. in a graft with two *crenulata* nuclei and one *mediterranea* nucleus the cap would tend to be of the *A. crenulata* type. The morphogenetic material is therefore species specific. It can be destroyed by ultra-violet radiation and inhibited by red light. In the latter case there is no destruction because the effect can be reversed by exposure to blue light.

It now appears that there are two morphogenetic substances; one is concerned with the initiation of cap formation and the other with controlling its actual shape. It is only this latter that is species specific. More recent work has suggested that there are one or more proteins involved in cap formation as well as a set of enzymes, both being separate from the proteins and enzymes concerned in stalk development.

When cap formation was studied in relation to light intensity it was concluded that the precursors of the morphogenetic substances can be formed under nuclear control at quite low light intensities. They are then activated by a higher light intensity which is not under nuclear control. At low values, e.g. 70 ft-candles, changes of intensity are more important than duration, while at higher values 225 ft-candles duration is more important. The amount of the morphogenetic substance produced seems to be proportional to nucleus size, the larger the nucleus the greater the amount of substance.

The importance of the nucleus in producing these morphogenetic substances has directed attention to protein synthesis. In enucleate plants RNA synthesis ceases, though protein manufacture can continue for up to three weeks. In decapitated rhizoidal portions the synthesis of RNA bears a close relationship to rate of regeneration of stalk and cap. When enucleate basal parts of *A. mediterranea* are provided with a hew nucleus, RNA synthesis takes place in subsequent new growth and not in the old stalk. Since RNA synthesis occurs in the plastids of enucleate segments it is clear that there are two sites for RNA synthesis and it is only the cytoplasmic one that is under nuclear control.

Growth in *Acetabularia* requires not only protein synthesis but also the morphogenetic substances. Because enzymes may be concerned with production of the morphogenetic substances attention has, in recent years, been turned to them. After enucleation invertase, acid phosphatase and phosphorylase all increase, and there is no lack of ribonuclease. Using interspecific grafts Keck found that acid phosphatase exists in two species specific forms, med and acic *(Acicularia schenkif)*, and that both are produced under enucleate conditions.

When med nuclei are transplanted to *Acicularia* stalks the acic phosphatase is replaced by med phosphatase within six days, though if no nucleus is introduced the acic phosphatase remains detectable for up to 25 days after enucleation. When an acic nucleus was transferred to a med stalk no acic phosphatase was subsequently produced. This may mean that production of med phosphatase is genetically dominant to that of acic. Further experiments showed that so long as med cytoplasm was present only med phosphatase was produced. It seems evident, therefore, that both nuclear and cytoplasmic factors may be involved in protein synthesis through species specific enzymes.

Up to the present there is no clue as to the chemical nature of the morphogenetic substances, but they can scarcely be I.A.A. or related compounds since these, although influencing growth in *Acetabularia,* are as a rule, primarily concerned in cell-wall extension.

CIRCADIAN RHYTHM

In recent years, there has been a lot of interest in the study of biological 'clocks' or circadian rhythm. A number of such rhythms have been known for a long time among the algae, the older ones being primarily concerned with lunar rhythms and reproductive-body discharge, e.g. Dictyotales, *Oedogonium* and *Halicystis.* In *Hydro-dictyon* an endogenous photosynthetic and respiratory rhythm operates during normal alternating light and dark periods and this continues when the cultures are transferred to continuous light.

The rhythm is also induced in red or blue light but not in green light. There is a comparable and even more pronounced rhythm in *Acetabularia* which continues for up to 40 days in enucleated plants, so that the phenomenon in this plant is not wholly under nuclear control. Similar rhythmic respiratory activity has been reported for *Fucus* though in this genus there has been no report of a photosynthetic rhythm.

Certain dinoflagellates (e.g. *Gonyaulax* spp.) are able to emit a flash of blue light resulting from the luciferaseatalysed oxidation of the substrate luciferin. When grown in dim light *Gonyaulax* demonstrates a diurnal rhythm in luminescence which is absent from those grown in bright light. The degree of luminescence is greatest at night. *Noctiluca* however does not show rhythm in luminescence. Other diurnal rhythms have been demonstrated for cell division in dinoflagellates, mitosis in *Spirogyra* and *Oedogonium*.

Photo-tactic tidal rhythms are also known. *Euglena Umosa* is positively phototactic but prior to high tide it burrows into the mud. The diatom *Hantzschia amphioxus* comes to the mud surface during low tide. These rhythms all appear to be temperature independent, but are affected by light. They are endogenous rhythms since they are maintained for some time after being brought into the laboratory and constant conditions.

In *Gonyaulax* the gradual decrease in the rhythm is due to energy exhaustion as *Gonyaulax* is an holophytic organism. This particular rhythm does not depend on prior growth under alternating light and dark conditions, because cultures grown in continuous bright light for a considerable period exhibit this diurnal periodicity.

PHOTOTAXIS, PHOTOTROPISM

Phototropic phenomena are known to occur in *Bryopsis, Acetabularia* and *Derbesia* while rather weaker effects are found in *Cladophora* and *Griffithsia.* In *Bryopsis* the tips of branches grow towards the light and rhizoids away from it and this may well be associated with the accumulation of auxin in the basal region.

Phototaxis is the movement towards (positive) or away from (negative) an external source of light. The response is one of two types: topo-taxis is movement towards or away from the light source while photo-taxis is the movement induced by a sudden change in light intensity such as occurs at the boundary between light and dark in a beam of light.

The Cyanophyceae show both photo- and topo-taxis with both positive and negative responses, depending on the light intensity. In *Phormidium uncinatum* the photoreceptor pigments may be both carotenoids and phy-coerythrin. Tactic responses are manifest in numerous flagellate groups such as the Euglenophyceae, Pyrrhophyta, Volvocales, *Ulva* gametes, Crypto-phyceae, Chloromonadophyceae.

Again both types of taxes and responses are known, and at least in *Platymonas subcordiformis* the response is dependent upon the external conditions. Low light intensity induces a positive taxis, high light intensity a negative taxis. The response is also dependent upon the light spectrum used, suggesting that two different mechanisms are involved.

In *Euglena gracilis* the phototactic response shows a daily periodicity with a peak reaction around mid-day. This response continues even when the *Euglena* is transferred to continuous darkness.

The photoreceptor is not the eyespot. Flagellates which lack eyespots are still phototactic. The receptor is believed to be located near the flagellar base, and the eyespot is thought to play a role in shading. Diatoms of the order Pennales are also phototactic. The responses are not always clearly defined, being influenced by light intensity and light quality. The irregular motion of the gliding diatom further confuses the interpretation.

The nature of the photoreceptor pigment is unknown, but action spectra studies indicate differences from class to class, and within the dinoflagellates, differences from genera to genera.

12
Ecology of Algae

ECOLOGY OF ROCKY COASTS

An observant investigator of rocky coasts cannot help but be impressed by the remarkably distinct zoning of the principal organisms; this zonation, moreover, is not confined to any one region but is found throughout the world. The vertical extent of the zones or belts depends on the tidal rise, the greater the rise the more extensive each zone.

Even in the Caribbean, where the tidal range is about nine inches, it is possible to observe a zonation. Any one of the major species commonly occupies a very definite vertical range, but occasionally may be found outside this range, the presence of a rock pool or cranny provides conditions favourable for its existence.

The numerous ecological studies that have now been carried out in many parts of the world have involved a great number of investigators. In any one area the number of communities recognized depends upon two factors.

- **The first factor to be considered is the locality.**

Some localities, especially those in colder waters, are very much richer than others. Within a region itself a rocky shore is usually richer than a boulder shore, and a protected area is commonly richer than an exposed one.

- **The second factor is personal analysis.**

Each investigator will tend to have a somewhat different concept of what comprises an algal community. The number of communities recognized will also depend upon the time and thoroughness with which the shore is examined.

The terminology that has been employed for naming the communities has led to much confusion. Algal ecology commenced after terrestrial ecology and so some investigators have attempted to apply terms used in the latter to the former, whereas others have considered that the conditions are sufficiently different to make this application impossible. Cotton recognised five algal formations, subdivided into associations, at Clare Island in western Ireland:

1. Rocky shore formation,
2. Sand and sandy mud formation,
3. Salt marsh formation,
4. River mouth formation,
5. Brackish bay format ion.

At Lough Ine in south-west Ireland, Rees recognized only two formations, the exposed and sheltered coast. Rees used the term 'association' for communities where species that are associated with the dominants are controlled by the same factors, but it is not easy to prove experimentally that certain factors do control the species distribution.

The term 'zone' was used for algal belts which possess horizontal continuity with well-marked upper and lower limits. The term zone has, however, a geographical conno'ation, e.g. temperate zone, and therefore the word 'belt' is best used for the marked horizontal communities.

In New Zealand, Cranwell and Moore used the term 'association complex' for the group of successive belts that follow one another vertically The horizontal belts are normally continuous though they can be interrupted occasionally by another community. In other words, there could be an association fragment of *Durvillea* in the *Xiphophora* belt.

There is still divergence of opinion about nomenclature, and even at the present time when it is desirable to secure uniformity, it is still perhaps more satisfactory to use the non-commital term 'community'.

It is important to have a satisfactory definition of the littoral. It can, be described in purely physical terms and is thus regarded as extending from mean high water (M.H.W.) to mean low watermark (M.L.W.), or from extreme high water (E.H.W.) to extreme low water (E.L.W.).

Tides are by no means uniform and the various irregular tidal types would lead to some difficulties. It is now realised that the animal and plant belts of the shore vary according to exposure, latitude and topography and can be to some extent independent of the tide. The present procedure therefore is to employ a biological definition of the littoral based upon the major organisms. Ultimately, when more is known about the causal relationships between tidal phenomena and the major belt organisms it would probably be more desirable to use tidal data. The biological criteria are essentially based upon the work of the Stephensons.

BASIC ZONATION

Stephensons proposed a basic grouping in the littoral which is now becoming generally accepted. Three major belts, the supra-littoral fringe, the mid-littoral and the sub (or infra-) littoral fringe, appear to be practically universal. The relationship of these belts to tide levels and the dominant organisms is as follows:

Supra-littoral	Upper limit for littorinids Extreme high water springs
Supra-littoral fringe	Upper limit for barnacles
Mid-littoral	Mean low water neaps Upper limit for laminarians or certain fucoids
Sub-littoral fringe	Extreme low water springs ? Exact position not settled
Sub-littoral	

This basic zonation varies from place to place depending on the interrelationships of tides and wave action. Some discussion has recently taken place about the nomenclature of these belts, but so many workers have accepted the Stephenson terminology that there seems as yet no valid reason for departing from it.

Lewis proposed that the upper limit of the littoral should be set by the upper limit of either the littorinid belt or of the black lichen *(Verrucaria)* – Cyanophyceae belt that replaces littorinids in some parts of the world. He sets the lower limit as the boundary between the Laminarians (or certain fucoids in the southern hemisphere) and the barnacle belt, and he does not recognise the sub-littoral fringe.

Other workers deny the existence of the sub-littoral fringe in their regions and therefore do not include it in a general scheme. Where it is recognised the transition is determined by the upper limit of the Laminarians, primarily in the northern hemisphere, or of fucoids, e.g. *Durvillea, Carpophyllum, Cystophora* in the southern hemisphere. The lower limit of this fringe is set a few feet below extreme low-water mark of spring tides, but at present our knowledge of this fringe is not as good as that of our knowledge of the mid-littoral and supra-littoral fringe.

Because it is submerged for so much of the time it is not an easy zone in which to work, and it is clear that more study will be necessary before the lower limit of this fringe can be satisfactorily settled. Although contributions to its flora and fauna are made by denizens of both the eulittoral and sub-littoral there are species which appear to be restricted to the belt. It seems to present particular difficulties in tropical regions as Lawson has pointed out in his study of the West African coast.

In most regions the supra-littoral fringe is characterised by the presence of species of*Littorina* or of allied genera, i.e. *Melaraphe* in New Zealand and Australia. As well as the littorines and, in some places, in their absence, the region is commonly occupied by Cyanophyceae and marine lichens and these form a conspicuous black belt. The transition to the major belt on the shore, the eulittoral, is determined by the upper limit of the barnacles, i.e. species of *Balanus, Chamaesipho, Chthalamus,* etc. In a given region, variation of exposure may determine which particular species sets the upper limit.

In most regions it is found necessary to subdivide this belt, such subdivision being based upon the dominant organisms. There is therefore considerable variation throughout the world> although some degree of uniformity is seen in the cold waters of the northern hemisphere on the one hand and the cold waters of the southern hemisphere on the other. At least a portion of the belt is normally dominated by the barnacles.

The levels of the upper belts in the eulittoral and of those in the supra-littoral fringe in any locality are not entirely dependent upon the height of the spring tides. On an exposed coast the shore is open to considerable wave action and a heavy spray rises against the rocks to a considerable height, and as a result both upper and lower limits are elevated several feet above normal. This elevation is known as the 'splash zone', but its extent depends on slope of shore and whether it is related only to storm waves or to fine-weather waves as well. On a very exposed coast with big wave$ the splash zone can be subdivided into,

1. The wash zone, which is the height the actual wave washes up the rock face at high water,
2. The splash zone, which is represented by actual splash from the wave,
3. The spray zone.

On exposed coasts the wash zone extends down the whole shore; the splash zone is generally only a few feet but the spray zone may be considerable. On the west coast of New Zealand littorinids occur up to 40 feet above the highest spring tides, and the spray from the giant waves can be seen drifting in nearly every day of the year. It is likely, however, that the size and growth of algal organisms in such elevated zones is not only due to the waves, but is also dependent upon moist air conditions that keep them damp during low tide periods.

Sufficient work has been carried out now to show that the bathymetric levels of the different belts are not uniform throughout a region although it could be argued that the controlling factors would be uniform. In Great Britain the bathymetric levels of *Fucus vesiculosus* and *F. serratus* vary from place to place. In New Zealand the belts of the major dominants show considerable variation, even after allowance has been made for splash elevation.

It has also been demonstrated that change of latitude may exert an influence upon the belts. Gislen has shown that on both sides of the Pacific the belts of brown algae tend to emerge with increase in latitude. Knight and Parke have shown the same thing to hold true for *Fucus serratus*', though work in New Zealand has shown that *Hormosira banksii* occupies significantly lower levels in regions further south.

Rock pools, couloirs (or gullies) and caves, represent other features on the rocky shore that afford rather specialized habitats. Rock pools have been divided into three groups:

1. Those of the upper eulittoral and supra-littoral fringe

which may remain exposed for several successive days during the neap tide periods and where therefore conditions of temperature and salinity can become extreme, particularly in summer;
2. The pools of the middle eulittoral,
3. Of the lower eulittoral, which may be covered continuously for several successive days during neap tides.

Algae found in the upper pools must be capable of tolerating rapid changes of salinity and temperature as well as being recipients of bird excrement. Such pools tend to be populated by mats of Cyanophyceae with which are associated diatoms and planktonic flagellates. Macroscopic algae are rare and are generally Chlorophyceae such as *Enteromorpha* and *Chaetomorpha.* On New Zealand shores *Pylaiella fulvescens* also occurs in such high pools.

Studies of the flora of rock pools must therefore take into consideration the levels of the pools and also the relation of the species to their occurrence elsewhere. This is necessary because the species can be divided as follows:

1. Species wholly confined to pools,
2. Species characteristic of the sub-littoral,
3. Adjacent eulittoral species,
4. Eulittoral species that reach an upper limit in pools that are above their normal limit on the exposed rocks. On exposed coasts deep pools can provide what is essentially a highly protected habitat and consequently the algae in such pools may be those of sheltered coasts rather than of exposed coasts.

The vegetation of couloirs or deep, narrow gullies is primarily affected by light conditions, since one or both sides commonly receive much less light than adjacent flat rock, and also by the heavy backward and forward surge of the tides. Specific algae may be associated with such couloirs,e.g. in New Zealand *Vidalia colensoi* is most frequent here; its presence may be due to the excessive water movement or to lack of competition from other algae in such a habitat.

In the case of caves the principal feature is the change in vegetation with decreasing light intensity. In a detailed study of the seasonal variations in light intensity for two New

Zealand caves Dellow and Cassie demonstrated that the pattern of zonation follows closely the gradient of indirect daylight, and that the species to be found could be divided into stenophotic or light-demanding species.

Peyssonelia, Hildenbrandia, Rhodochorton. The low light intensities found in caves favour what are fundamentally 'shade' species. In New Zealand the shade-inhabiting *Codium cranwellii* is restricted to caves in the littoral region, though it also occurs in the deep sublittoral where again light is reduced.

EUROPEAN ZONATION

In the North Sea and on the colder Atlantic shores of Europe the supra-littoral fringe is occupied by species of *Littorina* with which are commonly associated belts of *Porphyra umbilicalis* and *Pelvetia canaliculata.* Where the rock surface is extensive there is often a black lichen belt composed of *Lichina confinis* and *Verrncaria maura* above the *Littorina.* The *Peftetia* is more characteristic of sheltered shores and the *Porphyra* and *Lichina* of exposed shores.

The eulittoral is characterised by the presence of barnacles and a number of fucoid species. There are two main types of zonation depending on whether the dominant barnacles are *Balanus* or *Chthalamus* but transitions between these two types occur. It appears that *Chthalamus* can rise to slightly higher levels on the shore than *Balanus* so that the upper limit of the eulittoral can change its level from one place to another. Also, since *Pelvetia* normally grows above the *Balanus* but below the upper limit of *Chthalamus,* it may either belong to the littoral fringe or to the eulittoral.

Where the shores are not fully exposed the barnacles are associated with a succession of fucoid belts:

1. A belt of either *Fucus spiralis* or its var. *platycarpus.*
2. A belt of *Ascophyllum nodosum* with one of *F. vesiculosus* immediately below (or *vice versa)* with an intervening middle belt where both species occur.
3. The lowest belt of the mid-littoral is commonly dominated by *F serratus,* with which is often associated *Himanthalia,* or by species of*Laurencia* and *Rhodymenia* where there is shelter, or *Gigartina* or *Chondrus crispus* where there is more exposure.

One feature of the belts, in this part of the world, is the important part played by the fucoids. A comparable covering of algae is commonly absent in the corresponding cold waters of the Pacific.

Where the substrate is suitable, the sublittoral fringe is characterized by species of *Laminaria*, especially *L. digitata* but on exposed coasts *Alaria esculenta* is more characteristic. It is evident that the relative proportions of the different species vary with depth and locality and there may also be considerable variations. In the sublittoral when rock gives way to stones and then to shingle so the algal dominants change from *Laminaria digitata* to *L. saccharina* to *Halidrys* and finally to *Chorda* and small red and brown algae. A more detailed account of the belts can be found in Lewis.

On the rocky shores throughout the world communities of marine diatoms can be found. These, because of the taxonomic problems involved, have been but little studied. Aleem has shown that the diatom communities differ depending on the nature of the substrate, and there is also a difference depending on the actual bathymetric level. The different diatom communities are not to be found all the year round, but in general are at their best sometime between January to March or in the autumn and winter.

Fucoids and laminarians do not play so important a part in this region but they are replaced by other algae. In the far north, Nova Scbtia and Prince Edward Island the supra-littoral fringe is narrow and characterized by littorinids and Cyanophyceae, mainly *Rivularia*. In the eulittoral *Fucus vesiculosus* is the most plentiful fucoid but at low water it is possible to find great forests of *Chordariaflagelliformis* and *Scytosiphon lomentaria* in sheltered areas. In more exposed localities extensive areas of *Chondrus crispus* abound in the lower eulittoral and sublittoral. The Laminarian zone is not well represented on Prince Edward Island but is present on the Atlantic coast of Nova Scotia.

The belts in the North Pacific are quite different from those in the North Atlantic. The eulittoral is covered by barnacles but in the middle part there is a belt of *Fucus distichus* in one of its forms. The Laminarian zone is variably developed and in some places it is replaced by a mixed, rather impoverished algal growth. On the Oregon coast there is a well-marked series of belts commencing at the top of the eulittoral with *Pelvetiopsis fasligiata* or *Porphyra perforata* or *Scytosiphon lomentaria* and succeeded by *Fucus gardneri*. Species of *Cladophora* are also abundant with patches of *Cumagloia andersonii*.

The next major belt is dominated by species of *Iridophycus, Gigartina* and *Hymenena* and towards the bottom of the eulittoral it is replaced by a belt of the laminarians, *Hedophyllum sessile* and *Alaria*. In the sublittoral the dominants vary depending on local conditions and can be species of *Laminaria*^ *Lessoniopsis, Egregia menziesii* or *Nereo-cystis*. In sheltered harbours, e.g. Coos Bay, this belt carries beds of *Des-marestia* and *Laminaria saccharina*. In both littoral and sublittoral the marine phanerogam, *Phyllospadix,* occupies considerable areas.

In the southern part of California marine invertebrates are conspicuous in the upper belts of *Ralfsia—Prasiola, Endocladia — Gigartina* and *Halosaccion*. The lowest belts are dominated by *Laurencia, Sargassum* and *Cystoseira osmundacea* and at greater depths by *Egregia laevigata* and *Macrocystis.*

On the northern coasts the uppermost belt is dominated by *Pelvetia wrightii,* and the lower one by *Fucus evanescens* and *Laminaria longissima* while in the sublittoral there are large beds of Laminarians. In the drift-ice area on the Okhotsk coast a special belt dominated by *Heterochordaria abietina* and *Chordariaflagelliformis* occurs. In the more temperate waters, and therefore over most of Japan, there is an upper belt of *Gloiopeltis* or, more commonly, of *Myelophycus* and *Ishige* followed by an *Hijikia— Eisenia bicyclis* belt or a *Gigartina intermedia — Sargassum* belt where salinities are lower, i.e. where river waters discharge.

In the southern subtropical waters there are belts of small turf-like algae in which *Gelidium pusillum* and *Coraltina pilulifera* are dominants. Excessive wave action on the open coast eliminates *Sargassum* and favours the development of *Hijikia* and *Corallina*. The numerous inland waters of Japan contain a characteristic flora that differs from the open coast. At the higher salinities *Monostromay Scytosiphon lomentaria* and *Enteromorpha compressa* predominate, **whereas in less saline** waters a number of associations have been recognised in which *Ulva pertusa* and *Grateloupia filicina* or *G. turutusu* are the most widespread species.

SOUTHERN SEAS

In the South Pacific one of the principal features is that the major belts are dominated by animals. One reason for this is that, apart from *Hormosira,* the predominant fucoids are all genera that require total or almost total submergence, only being really exposed at low water of spring tides

Sarcophycus and *Durvillea* are found in the most exposed situations, while *Ecklonia, Hormosira* and *Cystophora* are typical of the more sheltered localities. *Xiphophora* and *Lessonia* occupy regions of moderate exposure.

In New Zealand a limited high-water belt of the tufty *Gelidiwn caula-cantheum, Bostrychia arbuscula* and *Enteromorpha* occurs in sheltered areas, whilst on exposed coasts there is a seasonal belt of *Porphyra columbina* var. *laingii*.

The *Hormosira* – *Corallina* – *Leathesia* belt is quite characteristic of the lower eulittoral on sheltered coasts and is replaced on exposed coasts by a belt of *Apophloea* and *Gigartina alveata.* The sublittoral species all have their upper limit above extreme low-water mark but none can tolerate any length of exposure.

Cystophora torulosa and *Carpophyllum maschalo-carpum* occupy only a narrow vertical range and are the characteristic algae of the sublittoral fringe. In the very protected waters of the New Zealand southern fiords the uppermost belt is dominated by the lichen *Verrucaria maura* and below it is a belt of the tufty *Bostrychia vaga* that is replaced by the typical *B. arbuscula* with increasing exposure. The *Hormosira banksii* belt is replaced by one dominated by *Cladophoropsis lyallii* and associated green algae.

WARM TEMPERATE ZONATIONS

In North Carolina only the summer flora is warm temperate, the winter flora being cold temperate. A similar combination of floras exists in the Chatham Islands off New Zealand. Along the Florida-Carolina coastline the supra-littoral fringe is represented by a 'black zone' dominated by Cyanophyceae and marine lichens.

The upper part of the eulittoral is dominated by barnacles *(Chthalamus, Tetraclitd)*, while the belt below by oyster and *Mytilus*. In this respect the warm-water coast here is similar to the warm-water east coast of the Auckland Province in New

The lowest belt on these U.S. shores is algal and dominated by species of *GracHaria, Ulva* and *Enteromorpha*. The sub-littoral fringe is represented by *Grateloupia* and *Sargassum*. The Stephensons have suggested that a mean winter water temperature of 10° C should form the limit between cold temperate and warm temperate shores. Bennett and Pope in Australia, however, regard the belts on the shores of Victoria as typically cold temperate, but there the mean winter temperature of the water is 11-8° C. In the event it may prove impossible to set an exact temperature at which the transition takes place.

On the Indian sub-continent the following belts appear to be generally distributed on the west coast, though the composition can vary with season.

1. Supra-littoral fringe: a belt of black Cyanophyceae with species of *Lyngbya, Pleurocapsa* and *Calothrix* as the dominants. Shallow pools in this belt contain several species *ofEctocarpus*.
2. Eulittoral: the upper part of this region is characterized by a bright- green belt of *Ulva* and *Enteromorpha* which may be mixed or may form a pure consociation of a single species. Shallow pools contain *Hypnea valentiae, Laurencia obtusa* and *Catenella repens*, while deep pools have *H. musciformis, Acanthophora delilei, Laurencia hypnoides* and *Caulerpa scalpelliformis*.

   The lower part of the region has a belt of *Colpomenia – lyengaria* and at the lowest levels a belt of *Gelidium – Polysiphonia – Ceramium*. The exact composition of each belt varies from place to place. In deep pools brown algae such as *Cystophyllum, Sargassum, Dictyopteris, Spatoglossum* and *Padina* abound and in shallow pools there is a green vegetation *of Caulerpa, Halimeda, Udotea* and *Cladophoropsis*.

3. Sub-littoral – this is dominated by species of *Champia, Codium, Caulerpa,* and *Chondria* and it exists in a number of associations.

In the warm waters of the world the sub-littoral is of greater importance. On the shore, exposure even for a relatively short time involves considerable desiccation and few species seem able to tolerate it. The sub-littoral flora depends primarily upon the type of substrate, whether sand, rock or coral.

Workers in the Mediterranean believe that with those clear waters a number of different belts can be recognized. Even in this sea there is a pronounced difference in the summer and winter aspects of the flora. Boreal Atlantic species such as *Ulothrix flacca, U. subflaccida, Bangia fusco-purpurea* and *Porphyra* spp. dominate the flora in winter, while in summer it is the tropical and sub-tropical species such as *Siphonocladus pusillus, Acetabularia mediterranean Pseudohryopsis myura, Liagora viscida,* which form the dominant species.

The zonations to be observed in South Africa are especially interesting. On the west coast the waters are relatively cold and a cold water flora and fauna predominate. The east coast is bathed by the warm waters of the Indian ocean and a very different flora and fauna exists. In the intermediate zone of the Cape itself there are transitions from the cold water zonation to the warm water zonation. Similar changes could no doubt be observed north and south of Cape Lookout in North Carolina and in other parts of the world. Very few of these changes have, as yet, been worked out.

In the very cold waters of the world, species are present that do not occur elsewhere. In the Arctic there are laminarians restricted to the region and in the Antarctic there are the peculiar genera *Himantothallus* and *Phyllogigas.* Nearly all the algae are sublittoral because ice action prevents littoral algae from establishing themselves. Subantarctic zonations e.g. Macquarie Island, Kerguelen, New Amsterdam, show an uppermost *Porphyra* belt and a lower eulittoral belt of *Rhodymenia,* with *Durvillea antarctica* (with Lithothamnia) the dominant of the sublittoral fringe in which *Ascoseira* and *Phaeurus* also occur.

In the sublittoral *Delesseria, Iridaea* and *Desmarestia* are dominants with forests of *Macrocystis* or *Lessonia* in deeper waters. Where *Phyllogigas* is present it occurs as a member of the lower sub-littoral, but even that is succeeded at lower depths by *Phycodrys* and *Phyllophora*. In the Arctic the uppermost belt is a black *Verrucaria* lichen with a lower seasonal belt of stunted *Enteromorpha compressa*. This paucity is related to the winter scour by ice. In the lower eulittoral, which escapes some ice action, *Fucus distichus* is a dominant in cracks and crevices. Below low-water mark a wider range of algae can be found associated with the laminarian belt.

Seasonal differences in the floras of different regions have been observed in various parts of the world. There are two general phenomena: one is the actual replacement of certain species by others, e.g. Mediterranean, North Carolina and elsewhere; the other is changes in the vertical levels of certain algae on the shore. Some algae migrate up in winter and down in summer, while others do the very reverse.

It is suggested that the nature of the response to temperature, controls the movements of those plants that migrate down in winter, and response to strong insolation determines the behaviour of those that move down in summer. It is likely that it is the response of the sporelings rather than the response of the adults that is involved, though the problem has not been investigated experimentally.

In the belts and zones that have been described so far, only what may be termed the normal has been considered. Variations in any region may occur as a result of changes in protection and exposure, e.g. in Europe *Ascophyllum* disappears in more exposed places and its niche is taken by *Himanthalia*. In New Zealand and Tasmania, *Hormosira, Ecklonia, Cystophora, Carpophyllum,* disappear on exposed coasts and are replaced by *Pachymenia* and *Durvillea* or *Sarcophycus*. Normally the substrate does not have a very profound effect on the main belts but chalk or limestone rocks do produce a significant change in the flora.

Fresh water, either in the form of substantial streams or rivulets induces a change in the flora. In the case of rivulets a dense growth of *Viva* and species of *Enteromorpha* marks their course through nearly all the normal belts on the shore. In the case of streams and rivers the dominants of the belts may change, i.e. in Europe *Fucus ceranoides* replaces *F. spiralis* in such places.

The relationship between animals and plants has not been as fully studied as it could be. There are, for instance, the algae associated with the hard shells of molluscs and barnacles. There are the browsing habits of limpets and sea-urchins.

Abundance of the former results in the absence of fucoids in the northern hemisphere, and the abundance of the latter results commonly in a complete clearance of algal vegetation. Pools heavily populated with urchins generally possess only coralline paint.

It should be evident that any really adequate study of marine coastal zonation must involve an appreciation of the part played not only by the algae but also by the animals. The ecologist must indeed be a biologist and the communities he recognizes must be biotic communities in the strictest sense of the term. The other alternative is for teams of botanists and zoologists to work together.

FRESHWATER ECOLOGY

One of the major problems in this branch of algal ecology appears to be the establishment of a successful classification on which field studies can be based. It would seem, however, that neither scheme alone is wholly satisfactory, but that a combination of the two provides a very suitable basis for workers in this field.

1. Subaerial communities
   These develop at their best in the tropics although they can also be found in temperate regions. The members form communities that are either epiphytic or epilithic.
2. Dripping rocks and waterfalls
3. Aquatic communities

These vary from season to season and frequently have a marked periodicity which is controlled by diverse factors. The following five subdivisions of the aquatic communities can be recognised:

a. Communities of rivers and rapids.
b. Spring communities.

c. Communities of bogs and swamps.
d. Communities of ponds and pools.
e. Lake communities.

SUBAERIAL COMMUNITIES

Aerial epiphytic algae are most abundant in places of high humidity and are particularly prolific in rain forest, i.e. tropical jungle, moss forest of the tropics and, to a lesser extent, in temperate rain forest. Species of *Cephaleuros* and *Phycopeltis* are common on leaves (Epiphyllophytes), *C. virescens* occurring as a pest on tea bushes, while species of *Trentepohlia* are also abundant, particularly on the trunks and branches (Epiphloeophytes) of various trees as well as on rock.

Diatoms can form a flora on mosses and liverworts and if the habitat is very wet Cyanophyceae and desmids may appear. Some algae, e.g. *Sticho-coccus bacillaris, Chlorella ellipsoidea,* have been recorded from the caps of Agarics (Fott, 1959); *Pleurococcus* is widespread on the trunks of trees but, together with *Prasiola* and *Trentepohlia,* is also a major component of the aerial epilithic communities (Lithophytes).

Certain algae seem to demand specific habitats, some of them unusual and among such algae are species of *Basicladia* that grow on turtles, and species of *Characium* and *Characiopsis* (all Epizoophytes) that grow on the Crustacean, *Branchipus.*

DRIPPING ROCKS AND WATERFALLS

Waterfalls have been very little studied. *Lemanea* frequently occurs in such habitats and there is often a rich flora of Cyanophyceae and Chloro-phyceae, particularly species of *Ulothrix, Lola,* and *Rhizoclonium.* Rock surfaces receiving seepage water from soils form a very suitable habitat for *Chroococcus, Gloeocapsa, Stigonema* and diatoms, including the colonial forms.

AQUATIC COMMUNITIES

Communities of Rivers and Rapids

In rapidly flowing rivers the algae are adapted to the habitat by either being encrusting, or, if they are only basally attached, the thallus is flexible and streaming. In such rapid streams attachment of motile reproductive bodies must be quick and efficient, though this aspect has not been investigated.

In rapidly flowing streams the flora is mainly composed of freshwater Rhodophyceae, e.g. *Lemanea, Compsopogon, Sacheria, Thorea, Hildenbrandia Cm* acid waters, *Batrochospermum* and *Bostrychia), Cladophora* spp. (especially *C. glomerata), Vaucheria* spp., Cyanophyceae, Chrysophyceae (especially *Hydrurus)* and diatoms. Very few of the species form pure stands, though *Hildenbrandia Heribaudiella (Lithoderma) Chaetophora* and *Rivularia* may form small compact communities.

In Europe *Hildenbrandia* frequently occurs with the lichen *Verrucaria rheithropila* (Rheithropilo-Hildenbrandietum). The mountain streams feeding the Ruhr River have been divided into two regions by Budde, an upper *Hildenbrandia* region and a lower *Lemanea* one. The epilithic diatom communities vary considerably. *Achnanlhes, Gomphonema* and *Synedra* are associated with nutrient deficient acidic waters, whilst in richer, more alkaline waters *Diatoma, Melosira varians* and *Meridian circulare* predominate. In less swift waters of larger rivers *Tabellaria* and *Fragilaria* may be dominant.

In European streams that are more sluggish *Enteromorpha* and *Hydro-dictyon* tend to replace *Cladophora,* and in very sluggish streams and also in lakes one may find temporary floating mats of *Hydrodictyon, Enteromorpha, Oedogonium, Spirogyra* and *Rhizoclonium.* With slower water movement sediments are deposited e.g. sand, silt, organic matter, and epipelic algal communities develop on these sediments. Such communities have not been much studied but diatoms are predominant, though one also finds Cyano-phyceae, Chlorococcales, Euglenoids and Desmids. In examining such communities care has to be taken not to confuse the species with remains of the epilithic and epiphytic floras.

The dominant benthic alga of rivers in the North Temperate zone is *Cladophora glomerata.* In the presence of certain metallic ions this species is adversely affected and may then be replaced by *Stigeoclonium tenue, Spirogyra fluviatilis, Phormidium autumnale,* etc. In Scandinavian streams it is often replaced by *Zygnema* and *Vaucheria.*

One of the most favourable locations for algal growth is

immediately downstream of large rocks though the exact significance of this fact has not been worked out. As might be expected, seasonal variations occur in the vegetation. Thus in the Ruhr streams the spring period is characterized by dominance of diatoms with *Ulothrix* and *Hormidium* as subdominants. In the summer Chlorophyceae and Desmids predominate and in the winter *Ulothrix* and *Hormidium* reappear. The effect of bank vegetation (overhanging trees, reeds, etc.) may be profound, especially in the degree of light reduction.

The most important controlling factor is apparently temperature but the degree of aeration and chemistry of the water is also significant. The absence or near absence of calcium, may induce a Cyanophycean rich flora. Light intensity variations and oxygen concentration changes may bring about local modifications of the water.

The distribution of algae and higher aquatic plants can provide an indication of stream pollution. Highly polluted, oxygen deficient waters usually only contain bacteria and possibly Euglenoids. With slightly less pollution *Oscillatoria* and *Phormidium* will be found. In the Saline River of Michigan the entry of certain metallic ions caused the normal flora to be replaced by *Stigeoclonium tenue* and *Tetraspora*. Unpolluted oligosaprobic waters can carry a rich algal flora.

It is possible to classify the communities of rivers and streams in the following way:

1. Communities of vertical substrates:
   a. Those attached to boulders, rock, mud banks or roots.
   b. Epiphytic communities on phanerogams.
   c. Spray communities.
2. Communities of horizontal substrates:
   e.g. those attached to stones, sand or embedded in mud.
3. Free-living communities.

The epiphytic communities are to be found attached to phanerogams, Bryophytes and larger algae, especially those such as *Cladophora* which have a rough, non-mucilaginous wall. The principal components of the epiphytic flora appear to be Cyanophyceae, diatoms and filamentous species of the

Chaetophorales, such as *Aphanochaete* and *Endoderma*.

There is still considerable scope for further work on river ecology, both descriptive and experimental and the use of artificial streams with controlled conditions should be invaluable.

SPRING COMMUNITIES

There are three main groups of springs: cold springs, hot springs and saline springs. The first named are usually rich in calcium and bicarbonate ions and the flora is generally that of alkaliphilic diatoms. *Balrachospermum* is the only common macroscopic alga. Saline springs usually have a much richer flora because the brackish water supports both marine and freshwater species. Again, diatoms are predominant but marine species of *Enteromorpha, Ulothrix,* and Cyanophyceae are also recorded. Hot springs are associated with volcanic areas and the waters have a high concentration of dissolved salts.

Cyanophyceae are the major, and often the only constituents, the various species being capable of secreting carbonate of lime or silica to form brightly coloured rock masses such as travertine and sinter, the rate of deposition sometimes being as much as 1 -25-1 -5 mm in three days. The highest temperature recorded for water containing living plants *(Synechococcus lividus)* is 74° C.

The number of forms capable of living in hot springs is considerable, no fewer than 53 genera and 163 species being recorded from the thermal waters of Yellowstone National Park. Recently it has been shown that one species, *Masligocladus laminosus* is capable of fixing atmospheric nitrogen.

It is evident that areas differ because Castenholz reports only 6-8 species from Icelandic thermal waters and then in temperatures below those of other countries.

The flora of such springs varies according to the nature of the water, and in very acid springs the most characteristic species is the Rhodophycean *Cyanidium caldarium* and the diatom *Pinnularia braunii* var. *amphicephala*.

COMMUNITIES OF BOGS AND SWAMPS

These associations are commonly found in waters that

are acid, particularly those in *Sphagnum* high moor bogs. They are very mixed with little or no periodicity, probably because of the relatively uniform conditions. Zygnemataceae, desmids and diatoms are most frequent, the desmid element changing considerably with altitude, chemistry of the water and type of substrate.

COMMUNITIES OF PONDS AND POOLS

The flora exists under very varied conditions with a regular or irregular periodicity. In the temperate regions Chlorococcales, Zygnemataceae (dominant in spring) and diatoms (dominant in winter) form the chief elements with a phytoplankton of flagellates. There is usually not enough aeration for the larger filamentous forms such as Cladophoraceae. The substrate and fauna are also important factors in determining the type of vegetation. The flora of tropical ponds contrasts sharply with that of temperate regions as there is:

1. An excess of Cyanophyceae
2. A scarcity of *Vaucheria, Oedogonium,* Xanthophyceae and Ulotrichales
3. An abundance of filamentous desmids together with *Spirogyra.*

In America Transeau concluded that freshwater pond algae can be divided into seven classes based on abundance, duration and reproductive season.

Species of some groups may be present every year and there are other species that occur sporadically. It is only after many years that extended cyclic periodicity becomes evident. Bethge recorded a ten-year cycle of growth for the production of cells of *Asterionella.*

In a study of a Harpenden pond Fritsch and Rich distinguished four seasonal phases:

1. Winter phase with *Microspora, Eunotia* and epiphytic diatoms, and with *Ranunculus aquatilis* and *Callitriche* as the dominant phanerogams.
2. Spring phase dominated by Conjugates, *Oedogonium* and *Conferva,* with *Ranunculus aquatilis* as the most important phanerogam.

3. Summer phase with *Euglena,* desmids and *Anabaena* associated with a phanerogamic vegetation *ofLemna, Glyceria* and *Bidens.*
4. Sparse autumn phase with *Lyngbya* and *Trachelomonas* but without any dominant phanerogam.

The algal periodicity is thus reflected by a similar periodicity in the phanerogamic vegetation. Compared with a Bristol pool the general trend of periodicity is very similar; there is a winter phase characterized by a hardy filamentous form *(Cladophora* or *Microspord)* and diatoms, and a spring phase with Zygnemataceae and an autumn phase with Oscillatoriaceae. The summer phase in the two pools was very different, and could be ascribed to the greater drying up of the Harpenden pool during that period.

Although the spring phases are essentially similar to those of the Harpenden pool with either *Spirogyra* or *Oedogonium,* nevertheless there are great differences.

The flora of pools is dependent not only upon general climatic conditions, such as rainfall and insolation, but also upon irregular microclimatic factors, e.g. aeration in the body of water itself. In many species there is a profound relationship between the meterological data and the frequency of the flora, e.g. *Microspora* and the Pleurococcaceae depend on temperature, *Oedogonium* and *Hormidium* depend on sunshine.

The factors influencing the growth of aquatic algae are:
(1) Seasonal,
(2) Irregular,
(3) Correlated.

The first group which are very obvious and need not be detailed, are important in large bodies of water, but they tend to be masked by the other two groups in small bodies of water:

1. Irregular factors:
   a. Abnormal rainfall or drought.
   b. Abnormal sunshine.
   c. Abnormally high or low temperatures.
2. Correlated factors:
   a. Species depending on the enrichment of the water by decay of other members of the flora.
   b. Forms influenced in their development by

competition with others.

c. Forms influenced in their development by the presence of a suitable host, e.g. epiphytic forms.

A correlation can frequently be established between the amount of sunshine and the phenomenon of reproduction. This is in accordance with experimental work which has shown that reproduction is often initiated by the presence of bright light.

An unusual concentration of the salts in the water during a period of drought may, however, counteract the influence of sunshine. There may also be a correlation with the nitrate factor for some species. Thus *Volvox* can receive a severe check when the nitrate is high and only reproduces at times of low nitrate value. *Euglena,* if present, attains to a maximum soon after the nitrate minimum unless other factors intervene.

Some pools may be permanently rich in an ion, e.g. calcium and carbonate ions in limestone areas, and they will possess a flora determined at least, in part, by this enrichment. West was the first investigator to show that a rich desmid flora is associated with an ancient hard-rock substrate (Oligotrophic waters), but diatoms are abundant in younger, softer-rock areas or where there has been much silting with consequent solution of mineral salts (Eutrophic waters).

A third type is the Dystrophic water, which is found on moorlands, where desmids form the most abundant part of the flora. In the course of years Oligotrophic waters may also change into Dystrophic waters.

Originally there was thought to be a fairly sharp distinction between oligotrophic, eutrophic and dystrophic waters, but with the realization that one type can gradually change into another it is evident that any such distinctions cannot be sharp. A considerable body of data is now available on the physical and chemical environment in pools and lakes. Among these physical factors is oxygen concentration. In sheltered lakes as compared with open lakes there is an oxygen stratification which closely follows the bottom contours, while the influence of any rivers entering the lake together with the problem of periodic floods is yet a further factor.

Where there is a shallow littoral shore the communities

may be difficult to recognise unless there is a rocky substrate, in which case there may then be a zonation that is dependent upon changes of water level and wave action: this type of zonation has been observed in several continental lakes. In deeper waters the communities are more distinct because a zonation develops which is primarily maintained by the light intensity factor or by competition.

The type of eulittoral flora is dependent upon the degree of exposure to wave action, the water chemistry position, (high or low) on the eulittoral and the vertical range of the eulittoral. Such factors are indeed reminiscent of those operating on a rocky coast. In Alpine lakes, e.g. the Traunsee, the uppermost belt is dominated by the Cyanophyceae, *Gloeocapsa sanguined* status *alpinus, Scytonema crustaceum,* and *Chlorogloea microcystoides*. Nearer the mean water level there is a belt of *Scytonema myochrous, Calothrix parietina, C. fusca* and *Tolypothrix distorta* van *penicillata*. On Lake Maggiore, on the chalk rocks of the eastern side, the upper *Gloeocapsa* is associated with *Scytonema myochrous* and two of its forms while on the siliceous rocky shores is a community very comparable to the *Scytonema-Calothrix* community of mean water level above. The spray zone of Lake Windermere carries a similar flora, the *Tolypothrix distorta* dominant being accompanied also by *Pleuro-capsa fusca*. At 5-25 cm below mean low-water *Rivularia haematites* occurs in the Traunsee, though if the shore is bouldery other species of *Rivularia* appear instead.

At still lower levels *Schizothrix lacustris* and Zygnemataceae become associated with *Rivularia*. On Windermere *Ulothrix zonata,* diatoms and Cyanophyceae occur around the low water zone.

On the clean extended beaches of Lake Maggiore the predominant species is *Coleochaete scutata,* though other Chaetophorales, e.g. *Stigeoclonium tenue, Chaetophora incrassata, Gongrosira de baryana* are very frequent. Below low-water mark the same algae are associated with *Dichothrix orginiana*$_9$ *Rivularia biasolettiana* and *Chaetophora elegans* and a very comparable community occurs on Lake Windermere at the same level. At intermediate depths in Lakes Windermere and Maggiore there are diatoms and filamentous green algae, while below 2 m Cyanophyceae once more predominate.

The epilithic flora of base-poor Swedish lakes comprises mixed Cyanophyceae above and below low-water level, but in the latter case the species present are associated with a rich diatom flora. Round has pointed out that there are three distinct components of lake floor algae. In protected bays of large lakes, e.g. Lakes Rotorua and Rotoiti in New Zealand, Chara-ceae may form thick underwater swards that have lasted for many years. In such places they have produced a repellant, black, reducing, organic mud of considerable depth.

In some European alpine lakes rounded balls of *Aegagro-pila* lie on the bottom and in oligotrophic lakes, it is possible to find in quiet bays, loose masses of *Spirogyra* and *Afpugeotia.Jhe* composition of the flora can be altered by the nature of the substrate, peat, sand, mud etc., and also by the sand-to-silt ratio. The composition of the flora also varies with depth and season.

An epiphytic flora develops on macroscopic phanerogams, bryophytes, lichens and large algae. This flora is influenced by depth, chemical composition of the water, season, nature of the host surface and life period of the host. On Lake Windermere Godward recognised the following epiphytic communities:

1. Epiphytic communities growing on aquatic macrophytes.
   a. On submerged plants between 0 and 0.5 m depth. This flora possesses a conspicuous Cholorophycean element, e.g. Conjugales, Chaetophorales and Ulothricales.
   b. On submerged plants between 1 and 3 m, dominated by *Oedogonium, Coleochaete* and diatoms.
   c. A community on submerged plants between 3 m and 6 m which is comprised of *Coleochaete,* a few diatoms and some Cyanophyceae.
2. Communities on dead leaves and organic debris:
   a. Up to 12m — wholly diatoms.
   b. Between 2-16 m — four diatom species and *Microcoleus delicatulus.*

The depth range of the diatoms was found to be greatest at the time of their maximum in spring and smallest in midwinter. It was also discovered that the diatom frequency and light intensity often show an opposite trend in the upper layers and a similar trend in the lower layers of the lake. The nature of the habitat, whether organic or inorganic, makes a considerable difference to the behaviour of the different species, and each individual species responds variously to the differences of these environments.

The nature of the substrate was also found to be of great importance. The tips of the leaves are concerned, the total number of epiphytes increases up to the third or fourth leaf from the apex, after which there is a decline.

The diatom flora however is an exception to this behaviour, because it increases regularly with the age of the substrate so that the oldest leaves bear the greatest number of diatomaceous epiphytes. There are distinct differences in the epiphytic flora of the upper and lower surfaces of the leaves, and it was observed that in the case of the first few leaves below the apex the upper surface was infinitely superior in the number of epiphytes, probably because of the greater light intensity.

In addition to distribution in relation to increasing age, there is also the relation to the different parts of the phanerogamic substrate. The distribution of epiphytes on the different parts of a phanerogam, and it will be observed that it is only on the leaf tips that the maximum is reached at the third or fourth leaf, while the leaf sheaths show a slight maximum at about the tenth leaf with a well-marked maximum for the mid-rib at the same level.

These maxima on the lower level leaves are to be associated with the diatom flora. It will also be observed that the number of epiphytes on the internodes remains more or less constant, but rapid growth of the substrate, e.g. the leaf lamina, tends to prevent colonization by epiphytes. The density of epiphytes that are attached to dead organic material is dependent upon the habitat of the substrate, e.g. if it is floating then there are few epiphytes, if it is attached or submerged the epiphytes are numerous, while if it is lying at the bottom the epiphytes will be few.

The various species to be found are all a residuum from the last living state of the material. In some cases the appearance of epiphytes is due to change in the host with age, e.g. old filaments of the Zygnemataceae lose their mucilage sheath and they then become colonized by many epiphytes.

When considering the effect of illumination not only are there problems associated with the individual plants, such as the upper and lower surfaces of leaves, but also the density of the host plants may be highly significant the distribution of various epiphytes on plants of *Equisetum limosum* under different conditions of spacing. Where there is screening of leaves, either on the same plant or by several plants, then the epiphytes develop on the unscreened portion.

The inter-relations of host and epiphyte are important, and epiphytes tend to develop in the depressions where the cells of the host adjoin each other. Results from experiments carried out with scratched slides suspended in the water make it clear that depression in a surface increase the number of epiphytes very considerably.

So far as the attachment organs of the epiphytes are concerned, there is no apparent relation between the nature of the substrate and the method of attachment. The behaviour of the motile reproductive bodies which either come to rest in the depressions or else are swept there by micro-currents in the water. Ponds with muddy bottoms have a reduced number of epiphytes probably because the pH and the gases evolved are toxic. Summing up, it can be said that the factors influencing the distribution of epiphytes are as follows:

1. Age of substrate
2. Rate of growth of substrate
3. Light intensity
4. Screening
5. Nature of the surface
6. Chemical surroundings.

Of these (3) is probably the most important, although it is difficult to separate the effects from those of (1) and (4).

PHYTOPLANKTON

Apart from attached littoral and sublittoral algae there is in the surface waters a vast population of floating algae (phytoplankton). This consists primarily of Bacillariophyceae,

Cryptophyceae, Pyrrhophyta, Cyanophyceae (particularly in fresh water), Haptophyceae, Chlorophyta (mainly in fresh waters) and Chrysophyceae, and the numbers are so enormous that it is in these waters that the rate of carbon fixation is very high. The growth of the crop depends primarily on five factors: rate of reproduction, rate of removal of individuals by death or grazing by zooplankton, photosynthesis, in relation to light intensity, availability of nutrients, and temperature.

To these it has recently been suggested should be added the effect of excreted algal substances, as they may partly determine the annual sequence of the organisms. Despite what may seem to be a prolific literature, as Strickland points out, our knowledge of primary marine and freshwater production is extremely meagre.

Plankton is found in the ocean, in freshwater lakes, ponds and rivers. It also occurs in abundance in specialized bodies of water such as sewage purification lagoons. In every case the flora is composed of larger microplankton species 50-500 m on diameter, which are readily collected in plankton nets, and smaller species 10-50 m diameter, the *nannoplankton,* which are only collected if water samples are filtered or centrifuged.

The principal component of the nannoplankton are Haptophyceae (Coccolitho-phorids), Silicoflagellates, Chrysophyceae, Cryptophyceae, Dinoflagellates, small flagellates and coccoid Chlorophyta. In many parts of the world there is little or no information about the composition of the nannoplankton. Some workers also recognize the ultra-plankton, organisms of 1-10 *m* diameter. There is no general agreement however about either the size divisions or the names.

Many of the species have elaborate morphological outgrowths but it is very doubtful whether these assist in flotation. Gas vacuoles and oil globules are probably the main buoyancy features, though if there is a mass of mucilage this may assist as its specific gravity is probably very close to that of water. However, the general consensus of opinion at present is that water turbulence is the main factor keeping phytoplankton afloat.

MARINE PHYTOPLANKTON

Vast areas of ocean are involved and for this reason the

geographical distribution of most species is only imperfectly known. Apart from variations in composition of the flora associated with the different water masses of the world, differences are also found between the coastal or neritic phytoplankton and the Open Ocean or oceanic plankton. The neritic plankton is commonly richer because the coastal waters contain more nutrients, and also because there are species derived from the non-planktonic littoral communities. Data is still meagre about the seasonal distribution of species and their occurrence at different depths. Variations in distribution are related to a number of different factors.

The primary factor is light, and the depth to which solar radiation can penetrate to permit positive net assimilation is known as the euphotic depth. Usually this is the depth to which 1 per cent of mid-day surface illumination penetrates in summer. This varies from one to two — in very highly productive waters rich in plankton, to 125 m in the clearest ocean waters.

The general range is between 15-75m. Another major factor is temperature; in Monterey Bay high temperatures favour *Ceratium* species and low temperatures favour *Thalassiosira.Thalassiosira antarctica* is a particular cold water species with a temperature tolerance between - 1 -8° C and 3-5° C and it also has a strict salinity tolerance being able to live in waters containing 32-6-34-5° salt.

The comparable Arctic species, *T. hyalina,* is likewise restricted to the colder waters of the Arctic and is absent where the warm Gulf Stream waters enter. In the Indian Ocean phytoplankton peak production is related to the south-west monsoon, which probably produces a great inflow of nutrients. It is also believed that the physiological state of the organisms, particularly as regards reproduction, is highly important.

Dinoflagellates are particularly valuable as a means of distinguishing various water masses in respect of chemistry and temperature, e.g. *Ceratium* species.vere used successfully in the Atlantic and Ferguson-Wood used the Dinoflagellates to delineate water masses around Australia. The North Sea has been studied very thoroughly with samples taken at known depths and within a very short space of time, and the various water masses delineated. A comparable study of the Ligurian sea revealed variations in the three dominant phytoplankton species in a transect across the sea and these variations probably reflect the existence of two separate water masses.

In recent years attempts have been made to use statistical methods, to establish ecological affinities in plankton populations, but it will take time before these methods become established to the extent they are in terrestrial ecology.

In nearly all temperate waters it has been found that there are distinct spring and autumn maxima when some component species occur in great abundance. In tropical waters there is very little variation, though in inshore waters, adjacent to coasts with monsoonal climates, a variation does occur associated with the rains.

In winter in temperate waters there is a rapid vertical mixing of the waters bringing up fresh nutrients from the lower layers, and this, when associated with increasing day length, more intense insolation and higher water temperature, is responsible for the development of the spring maximum. The spring blooms, particularly of diatoms in colder waters and in deep water, are further related to light intensity and light penetration. Continual vertical movement maintains a rich medium and, as a result, cold upwellings are very rich in plankton.

Lack of trace elements may reduce the numbers in certain species; iron affects the abundance of *costatwn,* and manganese and also vitamins, especially B_{12}, affects *Cyclotella* sp., as a result of their absorption by other organisms. The most spectacular of the seasonal maxima are the planktonic dinoflagellate blooms referred to as 'red tides', which are most commonly found in the temperate and tropical waters of California, Florida, South America and South West Africa. The two most common genera involved are *Gymnodinium* and *Gonyaulax* and sometimes *Prorocentrum.*

These blooms occur during the summer months and are closely related to the warm temperatures and abundant supply of nutrients caused by the upwelling of cold nutrient-rich water which is a common feature of these regions. Dinoflagellates are not responsible for all red tides, the red colouration in the Red Sea, is due to a bloom of the red Cyanophyte, *Trichodesmium*. Green discoloura-tions of the sea off the West coast of India have been caused by the chloromonad, *Hornellia*. The effect of upwelling water has been studied in detail in Monterey Bay where it causes high plankton volumes.

The steadiness of the upwelling is a major factor influencing the magnitude of local plankton blooms. These water masses have been studied in other parts of the world, the most intensive being in Walvis Bay, S.W. Africa. All upwelling areas seem to have a complex hydrography and this complicates the interpretation of biological data. Apart from the upwelling there is a relationship between the plankton volumes and presence of nutrients, such as phosphate and silicate.

An exception to the normal spring and autumn maxima is provided by the enclosed waters of the Mediterranean where the maxima occur in November-December and March, the one in March being the larger.

During the 1939-45 war, successful attempts were made to increase the plankton, and hence the fish, by adding artificial fertilizers to large, more or less enclosed, areas of water such as lochs. Generally only quantitative changes occur in the phytoplankton; qualitative changes develop only after a number of years unless there has been an algal bloom. A technique such as this can have valuable application, as, in the Philippine fish ponds.

In recent years new techniques, such as the continuous plankton sampler and the electronic counter have made plankton studies slightly less laborious. Data obtained from the North Sea using the electronic counter show distinct patches of Dinoflagellates. In the future more information is required about the seasonal stratification of phytoplankton, but this is related to the difficulties of occupying ocean stations for any great length of time, and is also related to the fact that at a fixed station the body of water itself can change. In the North Sea, where this kind of work has been done, changes only occur with depth in summer when a thermocline develops.

A vast amount of work has been performed on the productivity of oceanic waters with determinations of oxygen production using either Winkler light and dark bottles or determinations of the amount of radioactive ^{14}C tracer absorbed by the cells under given light conditions.

Discussion is still taking place about the validity of the various techniques associated with the use of ^{14}C, but the available results do give at least an indication of the productivity of various waters. Another technique is to determine the amount of chlorophyll *a* present in a sample of phytoplankton, and fair agreement has been reported between the two methods.

Danish workers have distinguished four classes of oceanic production in tropical and subtropical zones:

1. Daily productivity 0-5-3 g C produced/m^2 in one day. This is associated with the considerable admixture of nutrient-rich deep water into the surface waters, e.g. Benguela current.
2. 0-2-0-5 g C produced/m^2 in one day associated with the fairly steady mixing of nutrient-rich waters such as occurs with the Equatorial counter currents.
3. 0-1 g C produced/m^2 in one day where any mixing of nutrient-rich waters is slow.
4. 0-05 g C produced/m^2 in one day in regions of old, nutrient-poor waters, e.g. central Sargasso Sea.

On the whole, however, there are still area for which there is no information, particularly in the Pacific and Indian Oceans.

Plankton productivity can be compared with the productivity of epipelic algae which are mainly diatoms growing on mud and sand flats. On salt marshes in Georgia, Pomeroy found that at high water a mixture of diatoms and Cyanophyceae fixed the equivalent of 0-2 g C/m^2 over a 10 hour day, while at low water the rate was greatly reduced, only 0-02 g C/m^2 being fixed.

This compares with data from the mud and sand flats of the Danish Wadden sea where values ranging from 0-038 to 0-047 g C produced/m^2 over a 10 hour day have been recorded. Variations in productivity as between groups of algae can be very considerable: thus Cyanophyceae can fix 35-390 mg C/kg/h., a mixture of diatoms and flagellates produce 2-4-757 mg C/kg/h and pure diatom communities produce 4-454 mg C/kg/h.

In the Antarctic, studies of plankton, mainly *Phaeocystis*, below the ice, as well as of plankton from the bottom of the sea-ice itself, has shown that both the ice flora and the plankton have a degree of 'shade' adaptation well in excess of that of Arctic plankton. The ice organisms show an increasing inhibition of photosynthesis above 100 ft-candles, and light saturation with the antarctic plankton was reached at 300 ft-candles. Temperatures below 5° C brought a very great drop in the carbon fixation, and this is very significant as temperatures in the experimental waters rarely rise above -1-5°C.

The whole problem of nutrition in antarctic waters demands further study. In recent years some very interesting work has been done on the productivity of artificial marine systems. Artificial reef ponds were set up in Texas with suitable animals and plants and these developed heavy nannoplankton blooms. In such habitats careful measurement of the environmental factors can be maintained and these showed some resemblance to data obtained in the field. Comparable systems can be established in the laboratory to enable microenvironments associated with algae.

FRESHWATER PHYTOPLANKTON

Methods of collection are the same as for marine phytoplankton, and in both cases it is most desirable to examine the living material, even if subsequently there has to be preservation.

The components of freshwater phytoplankton differ from the marine in that there are a greater number of species of Cyanophyceae, abundant Desmids (completely lacking from salt water), Volvocales and Chloro-coccales but fewer Pyrrhophyta. In inland salt lakes a few marine forms can be found, but although genera such as *Peridinium* and *Ceratium* are common, the species may be specific to salt lakes.

The larger the lake the more chance there is of horizontal variation. Thus in the Great Lakes of North America there are at least three different communities in the western part of Lake Erie. These have been studied as to composition and photosynthetic yields. In the latter case there are considerable discrepancies between the different methods and this may be related to lake turbulence.

Apart fr6m this kind of phenomenon the plankton of inshore waters tends to be richer through admixture of benthic species. In lakes that develop thermal stratification the phytoplankton are effectively limited to the epilimnion because the water movement below this level is so slow the cells sink to the bottom. In large lakes the hypolimnion is often in the zone to which insufficient light penetrates for photosynthesis. In the case of very turbid eutrophic lakes, light penetration into even the epilimnion may be a limiting factor.

The effect of eutrophication of freshwater lakes on the phytoplankton is becoming of great importance because the ratio of desmids to other selected algal groups can be used as a measure of eutrophication. Despite this it is often very difficult to define the terms precisely as different investigators have used different approaches. There are also seasonal variations to be taken into consideration. A well established example of the eutrophication process is provided by the Great Lakes of the St. Lawrence basin.

From these figures it will be seen that Lake Michigan is on the verge of passing from one type to the other. The changes that have been taking place over the years, particularly in Lakes Ontario and Erie. In western Lake Erie the changes have been particularly pronounced since 1950 and these are reflected in the phytoplankton.

There it is associated also with excessive growth of the introduced aquatic phanerogam, *Lagarosiphon major*. This has reached nuisance proportions and is being presently controlled by diquat spraying. The elimination of the *Lagarosiphon* brings about some change in the phytoplankton which could be associated with the release of nutrients. This, however, is a feature that is being studied. It seems that in future forest buffer zones ought to be maintained around lakes to trap the run-off of surplus nutrients.

Less work has been performed on primary production in Lake Phytoplankton than in the sea. Using ^{14}C incorporation techniques primary production in Danish lakes was found to range from 300 to 370 g C m^2/yr, while in Swedish oligotrophic lakes carbon fixation is less than 0-1 g C m^2/yr though in eutropic lakes the value may be 2-2 g C m^2/yr, under optimal conditions.

In general, maximum absorption of carbon occurs at a depth below 0-5 m. In some waters there is a clear relationship between primary production and transparency, and between pH and primary production. When transparency decreases the phytoplankton crop increases and so does the pH.

In any large lake it is not sufficient to study phytoplankton from a single station or at one season. Sampling in a number of stations shows that variation occurs within lakes and there is also a seasonal variation related to nutrient supply, and in some lakes to the stratification cycle. Of the nutrients, nitrate and iron are most likely to limit algal growth.

Relatively little work has been done on the effect of temperature on carbon assimilation by phytoplankton. Some recent work by Aruga has shown a relationship between the optimum temperature for photosynthesis and the water temperature. In spring and winter the optimum temperature for the flora is around 20° C with actual temperatures of 8-15° C. In summer the optimum temperature and the actual water temperatures more or less coincide, reaching a maximum of 30° C in July and August.

The low temperature of high altitudes is obviously of paramount importance, but this factor has also to be considered in relation to the seasonal light intensity. The interaction of these two factors produces a plankton maximum immediately after the ice thaw and again in the autumn.

Compared with the limnoplankton of lakes our knowledge of the potamo-plankton of rivers is extremely meagre. Indeed it was not until nearly 1930 that the existence of a separate potamoplankton really became recognized.

Before then it was thought that river plankton originated from outside sources. Potamoplankton can be subdivided as follows:

1. Eupotamic, thriving in the stream and its backwaters.
2. Tychopotamic, thriving only in the backwaters.
3. Autopotamic, thriving only in the stream.

A potamoplankton survey of three rivers in the Goros region showed that there were three maxima for diatoms and Chlorophyceae, and four for Cyanophyceae. The total flora comprized 155 species. It seems, however, that even though there may be a rich phytoplankton the component species do not form a distinctive association. In general the development of potamoplankton is dependent on the age of the water and whether the stream is in flood or drying up. The plankton of small streams appears to arise from the benthos.

Apart from the normal numbers of planktonic forms present in river waters, one or more species can from time to time develop into a bloom, but this is related to rate of water flow. One interesting feature of such blooms is their restriction to relatively small stretches of a river. In the case of the Danube blooms they even remain in the same place for two or three days.

At present there is no satisfactory answer to this phenomenon. In the case of the Danube, which has been studied fairly thoroughly, there does not seem to be a definite plankton cycle or any specific maxima. One important feature of potamoplankton is the high proportion of diatoms present. Thus the following percentages have been recorded: Volga, 45-53%; Dnieper, 48%; Rhine, 56%; Vistula, 55-74%.

In recent years increasing use has been made of the unicellular green algae, especially *Chlorella, Scenedesmus* and *Chlamydomonas,* to purify sewage effluent. Most of the pioneer work has been carried out in the United States where large lakes of effluent are innoculated with the algae, and these after a few days increase in such number that the oxygen they produce aids the aerobic decomposition of the remaining organic matter. When one of these oxidation lakes is operating, water is allowed to flow through at a rate which takes it about 6-7 days to pass from entry to exit, or else it is allowed to recirculate.

At the end of this period the effluent emerges clear of organic matter and the bacterial count of pathogens has also been reduced to a negligible value. At present the largest oxidation lakes for sewage purification are at Auckland in New Zealand, where a detailed study of the phytoplankton has recently been carried out. A very substantial planktonic flora, especially in species of *Euglenp* was found.

Small lakes, such as those used in Israel, have a much more restricted flora and at times can be almost unialgal (e.g. *Prymnesiwn parvurri*). The dominant species of algae are substantially the same in ponds in all countries.

The large Auckland ponds, however, contained 36% of species not previously recorded in this habitat and there were two new species. A characteristic of the Auckland ponds was a paucity of Volvocales, especially *Chlamydomonas,* which are common elsewhere. The Volvocales are generally the first algae to develop in sewage treatment ponds but they are replaced later by *Chlorella* and *Scenedesmus.* There was a seasonal maximum in the warm weather which was probably related to the higher light intensities and temperatures, though there was also a heavier loading of the ponds. The flora was also more varied at this time.

Different types of treatment pond appear to carry different floras:

1. Very lightly loaded ponds or polluted natural waters develop blooms of blue-green algae *(Microcystis)* in summer and a mixed Chlorococcalean flora at other times.
2. Highly loaded ponds with low ammonia have a mixed Chlorococcalean flora with other types, e.g. *Closterium cornu, Cyclotella meneghiniana,* abundant at times.
3. Ponds with a low biological oxygen demand (B.O.D.) and high ammonia are characterized by euglenoids.
4. Ponds with a high B.O.D. have floras dominated by *Chlorella* and *Selenastrum miniatum,* with *Ankistrodesmus falcatus* var. *acicularis* or *Scenedesmus opoliensis* often becoming important.

In conclusion it is evident that much more work is needed in pure taxonomic phycology to find out which species are predominant in different areas and to catalogue, preferably statistically, their associations as well as the successions. The factors affecting the growth and distribution of phytoplankton are a complex of physical, e.g. light, temperature, turbidity, water movement; chemical, e.g. nutrients and excreted organic matter, and biological, of which grazing is perhaps the most important. The utilization of the nutrient varies with different species and is complicated. There is not present much data on the relationship between growth and concentration of nutrients, in particular nitrate and ammonia.

It has been known for some time that living planktonic cells, marine and freshwater, excrete organic matter, which may range from 0-5 to 34% of the total photosynthate. In many species the average amount is 3-6%. A few of the products are highly toxic, but we know very little about some of the other compounds. One important compound is glycollic acid.

This liberation of a large proportion of photosynthetic product into the medium, whether sea or freshwater suggests that the water always contains material that can act as a source of energy and can exert an ecological effect. Excretion takes place in the light as well as in the dark, but recent work suggests that the phenomena are different. Since there is a relationship with light intensity the amount excreted varies with water depth, increasing with depth and decreased light intensity.

The material excreted is not rich in carbohydrates, but may contain considerable quantities of amino acids and peptides, with lesser amounts of protein and chloroform – soluble substances. A few species excrete a single compound almost exclusively. There is also good evidence that some algae excrete substances that inhibit their own growth and that of other algae.

Finally, reference must be made to a peculiar type of freshwater plankton, the cryoplankton of the snow, e.g. red snow due to the presence of *Chlamy-domonas nivalis;* this is by far the most common species, though in the Pacific north west of America C. *yellowstonensis* is also very important. The colour is due to carotenoids in the cells, especially in the resting spores.

Algae can also give other colours to snow; yellow snow associated with a flora of about twelve species, all containing much fat; green snow, principally caused by zoogonidia of green algae; brown snow or ice due to the presence of *Mesotaenium* and mineral matter; black snow caused by *Scotiella nivalis* and *Rhaphidonema brevirostre;* and a light, brownish-purple ice-bloom caused by a species of *Ancyclonema.*

Cryoplankton is normally only present where the snow or ice is stable for a considerable period.

13
Reproduction in Algae

As a result of recent work on pigments in simpler members of the algae, together with electron microscope studies of cell structure, we are in a better position to assess the inter-relations of the classes than hitherto. Any evolutionary scheme, however, can only be regarded as tentative. The first photosynthetic organisms would have been those with such pigments as the bacteriochlorophylls or chlorobium chlorophylls that acquired the ability to synthesize magnesium tetrapyrroles. Such organisms would also have been anaerobes because oxygen would not have been evolved and the electron donor would have been a sulphur compound or an organic compound (e.g. organic acids).

With the development of chlorophyll and the evolution of Photosystem II (involving such pigments as biliproteins and secondary chlorophylls), the ubiquitous water became the electron donor and consequently oxygen was produced. With the production of oxygen came aerobiosis.

Probably the simplest of these plants were the blue-green algae, a stock that contained chlorophyll and biliproteins. The transition from these organisms to the simplest eucaryotic type is obscure; although it would appear that something like *Cyanidium caldarium* or the Cyanellae may represent such a transition form.

The whole problem involving the origin of chloroplasts and the relationship between Cyanophyceae, Rhodophyceae, Cryptophyceae and the Cyanellae has recently been discussed by Allsopp and Carr and Craig. The Rhodophyceae and Cryptophyceae are considered as primitive eucaryotes that were evolutionary dead ends. From the simple Eucaryotic ancestors, whatever they may have been, there appears to be two predominant lines of evolution.

The main line of evolution is associated with the origin of chlorophyll *b* and starch-type, storage polysaccharides; it forms the *ab* line with the Charophyceae, Chlorophyceae, and Prasinophyceae and from them the land plants. This assemblage has been called by Christensen the Chloro-phyta. The other main line of evolution is associated with the presence of chlorophyll *c* and laminarin-type storage polysaccharides; it can be termed the *ac* line. This group has been called the Chromo-phyta by Christensen.

In so far as the green algae or Chlorophyceae represent the line from which it is believed land plants ultimately evolved it is convenient to consider evolution within that group first. There have been two major views about the origin of the group. According to one the Chlamydomonadaceae are regarded as the primitive source, whereas according to the other the source is the Palmellaceae.

One line leads to the development of aggregate spherical colonies, culminating in *Volvox* another line leads to the production of gelatinous colonies, which may become net-like *(Schizodictyori)*. All these, however, would seem to represent 'blind alleys' in evolution. Studies of flagella and associated apparatus are contributing towards phylo-genetic understanding but at present they do not indicate any major deviation from the above.

The names of the genera do not necessarily imply that they formed-the actual intermediate stages, but merely that forms like them were possibly involved. In the main sequence of evolution the logical development would appear to be the non-motile unicell and then the attached unicell. From the former could have developed more colonial forms terminating in giant nets *(Hydro-dictyori)*, but simple siphonaceous forms may also have been produced *(Protosiphori)*.

The non-motile unicell would develop into various types of simple filament from whence evolution would appear to have progressed in three directions:

1. To parenchymatous types, e.g. *Ulva, Enteromorpha.*
2. To partially coenocytic types, e.g. Siphonocladales, and thence to the Siphonales. It has been suggested as an alternative that the Siphonales may have arisen from the siphonaceous Chlorococcales *(Protosiphon)*.

3. To branched heterotrichous types (Chaetophorales). This latter line, like certain Ulotrichales would also show heteromorphy in the life history. It is the line from which it is believed the terrestrial flora arose.

The Dasycladales with the large single nuclei seem to occupy a special position and it is difficult to support an argument that would locate them between the vesicular Siphonocladales and the Siphonales.

Latter aspect is more in line with Kornmann's suggestion that the Ulotrichales are an isolated group. It also differs in that the Chlorococcales represent the ancestral stock of the Siphonocladales, Siphonales and Dasycladales.

The ancestors of today's heteromorphic Ulotrichales probably possessed isomorphic alternation of generations from which arose the current paren-chymatous members with their isomorphic alternation. It is also probable that heteromorphy arose more than once in the course of evolution, e.g. Monostromataceae and Acrosiphoniaceae.

Furthermore it seems more probable that the Siphonales and Siphonocladales arose from such stock, rather than from the Chlorococcales (e.g., *Protosiphon* — like organism), for the following reasons:

1. There are no intermediates between the wholly haploid *Protosiphon* and the wholly diploid Siphonales although one might expect to find some forms with both generations alternating,
2. The Chlorococcales are nearly all freshwater whereas the Siphonales are all marine,
3. Intermediates can be envisaged from at least some members of the Siphonocladales and Siphonales.

A more detailed discussion of this problem will be found in Chapman and Nizamuddin. Recently Hori and Ueda have proposed two lines of evolutionary development within the Siphonales based upon plastid structure, but evidence based on cell ultra-structure alone must be treated with caution.

In the Siphocladalean line the position of *Urospora,* with macroscopic gametophyte and dwarf sporophytic, 'Codiolum' stage is of interest because it shows how forms with heteromorphic generations could have arisen, e.g. *Spongomorpha, Derbesia.* The other feature is the appearance of heterotrichy in the parenchymatous group, and also in the most advanced forms the disappearance of the haploid generation.

The concept of the heterotrichous habit was first advanced by Fritsch, and it is clear that an appreciation of this phenomenon is important in understanding evolutionary problems among the algae. After the intercalation of a sporophytic generation there is also the disappearance of the haploid generation associated with development of the vesicular forms of the Siphonocladales.

With regard to the Conjugates two viewpoints have existed:

1. The filamentous members arose from a Ulothricaceous stock and the unicellular forms developed from the filaments by reduction and fragmentation.
2. The nature of Ulothricaceous reproduction makes seem unlikely and instead the suggestions of recent workers have been followed, namely, that the unicellular Conjugates arose from a non-motile unicellular stock, and that these in turn gave rise to the filamentous forms.

Accepting this viewpoint the Conjugates represent two separate lines of evolution stemming from the unicellular Mesotaeniaceae. The one line gave rise to the unicellular Desmidiaceae with exozygospores and only a hint of the filamentous condition, as seen in *Desmidium* and *Gymnozyga,* and the other line saw the evolution of the filamentous Zygnemataceae with endozygospores. The filamentous genera probably represent three separate lines of evolution based upon the nature of the chloroplast, whether stellate, plate-like or ribbon-like, and each of these lines shows the same six parallel evolutionary trends in respect of differentiation of gametes and zygospores.

With regard to the Oedogoniales, one can only assume that they arose, with much modification, from a filamentous Ulothricaceous stock. The flagellate Prasinophyceae are considered, on grounds of their ultra-structure, as being primitive – perhaps the most primitive of all green chlorophytes. At this point, however, it is perhaps too early to speculate on their relationship to the primitive Chlorophyceae. At the present time it seems that the Chaetophorales and Chlorococcales are the orders most in need of further study.

In the Phaeophyceae, the primitive members (Ectocarpales) known today possess a thallus of branched heterotrichous filaments, with alternation of isomorphic generations. Biochemically the origin of the Phaeophyceae would appear to be with the Chrysophyceae. Both groups may have arisen from a common stock, or the Phaeophyceae may have had an origin within the Chrysophyceae via an intermediate form such as *Phaeothamnion.*

Within the Phaeophyceae themselves those members with isomorphic alternation of generations probably have some relationship with one another. The apparently difficult case of *Cutleria* and *Aglaozonia* can be explained on the basis of an original heterotrichous ancestor in which the aerial part of the gametophyte persisted and only the basal part of the sporophyte.

In the case *of Microzonia* one generation seems to have disappeared completely. Orders with heteromorphic alternation (Heterogeneratae) can conveniently be divided into those that are pseudoparenchymatous (Haplo-stichineae) and those which are truly parenchymatous (Polystichineae).

It has been suggested that the Fucales might be derived from the Chor-dariales (Mesogloiaceae). The principal reasons are associated with the presence of the apical hairs falling off and giving rise to the adult apical growing cell, because a similar type of behaviour is to be found in *Acrothrix;* in gross structure also, e.g. primary and secondary medullary filaments and the assimilatory tissue, there is some slight resemblance between that of *Fucus* and *Eudesme.*

The comparative looseness of the *Mesogloia* type of structure and the compactness of the Fucalean does, however, present difficulties. The suggestions involving *Chordaria, Acrothrix* and *Splachnidiwn* indicate that perhaps the Chordariales occupy a key position within the Phaeophyceae.

The importance of the Chordariales has been further stressed by the discovery that the former Fucalean species *Notheia anomala* reproduces by means of macro- and micro-swarmers and is not a member of the Fucales at all. Its affinities are probably not far from *Splachnidium*, which might well represent a relic of ancestral forms from which the Hormosiraceae were derived.

The adult plant is sporophytic and there is a dwarf generation which is presumably gametophytic though no reproductive organs have so far been observed. As has been pointed out, it may well be that in *Notheia* the male gametophyte is not produced and the micro-swarmers act as antherozoids. A dwarf filamentous female gametophyte is still produced but is apparently not liberated. Reduction of this phase would lead to the Fucalean condition found in *Hormosira*.

Formerly included in the Fucales, the Durvilleales could well have been evolved from a Laminarian complex. Recent interesting discoveries about reproduction in the peculiar antarctic genus *Phyllogigas* support this, as also has work by Naylor on *Durvillea*. The remaining Fucales are probably diphyletic, the Hormosiraceae being derived from a Chordarialean source such as *Notheia*, while the Cystoseiraceae and Sargassaceae may be derived from the Chor-dariaceae.

The above considerations also emphasize the great phylogenetic importance of the Dictyosiphonales and Chordariales. A detailed study of the former has still to be carried out, but since the Chordariales form a more or less homogeneous group, it is possible to suggest how evolution within the order may have taken place.

The Chordariaceae would seem to represent the primitive stock with uniaxial plants *(Mesogloid)* preceding multiaxial plants *(Eudesme, Myriogloia)*. At some stage reduction must have occurred leading to the reduced epiphytic Myrionemataceae and the saxicolous, crustaceous Ralf-siaceae. Another line of development is provided by the cushion forms such as *Elachista*. The final stage is probably represented by a genus such as *Corynophlaea* with *Petrospongium* being treated as a reduced form.

Yet another line of evolution must be represented by the pseudoparen-chymatous Spermatochnaceae with apical growth instead of the usual trichothallic. *Spermatochnus* and *Nemacystus* with a single axial thread may be regarded as more primitive than genera with several axial threads (e.g. *Stilophord).*

Pseudoparenchymatous development is also found in *Chordaria,* representing the most advanced condition in the Chordariales, and in the peculiar monotypic genus *Splachnidium.* Another peculiar monotypic genus with no immediate obvious affinities is the antarctic *Caepidium.* Here there is a basal thalloid plant, reminiscent of *Ralfsia,* from which at certain times arise branched threads of *Chordaria at* structure. The plant also produces a *Leathesia-like* cushion form.

RHODOPHYCEAE

The origin of the Rhodophyceae poses numerous problems. Among the Eucaryotic algae there appear to be no immediate relatives of this group. The class is completely lacking in any flagellated stage, and biochemical differences include the presence of large amounts of biliproteins, a simple carotenoid distribution, Floridean starch and a tendency to form polysaccharide-sulphate esters.

It has been suggested that the Rhodophyceae may have arisen from the Cyanophyceae. The main argument for this relationship is the similarity in pigments, especially biliproteins. It should be remembered, however, that any connection between the Rhodophyceae and Cyanophyceae would transcend the demarcation between the Eucaryotic and Procaryotic organization and all the inherent differences that go with this.

Recently Chadefaud has suggested another possible line of development based upon a study of the reproductive organs. He points out that no sex organs are known in the Cyanophyceae whereas they are present in primitive Rhodophyceae (other than Porphyridiales) and that there is also evidence of sex in certain bacteria; hybrids are known and a form of meiosis occurs at endospore formation in the Sporobacteriales in which the donor cell passes genetic material to the receptor cell.

This process can be termed archeogamy. In *Petrocelis cruenta* a nucleus is transferred from one filament to another by a connecting thread, though there is no subsequent nuclear fusion. In *Rhodochaete parvula* a nucleus is transferred from the spermatocyst by a spermatium to a receptor cell and the two nuclei fuse. This process can be regarded as protogamy. There are two possible variants from this condition.

In one the spermatocyst becomes terminal and attached by a stalk (Florideophycideae) while in the other the entire cell functions as the spermatocyst (Bangiales). Chadefaud argues that the same kind of sequence can be seen in the development of the carposporophyte. If this interpretation is correct we would have to regard the bacteria as probably giving rise separately to the Cyanophyceae and Rhodophyceae, and algae, such as *Porphyridium* and *Chroothece,* would then have to be regarded as reduction forms rather than primitive genera. There is a thermophilic, acidophilic organism, *Cyanidium caldarium^* which may be regarded as a very primitive Rhodophyte.

The Florideophycidae presumably arose from within the Bangiophycideae, though not necessarily from the most advanced members. Any phylogenetic treatment must take into consideration their sexual reproductive organs, and the auxiliary cell.

Little attention has been paid to problems of phylogeny within the Florideophycidae. Recently Fan has suggested that the Gelidiaceae and Cryptonemiales probably originated from a common ancestral stock which in its turn had been derived from members of the Nemalionales in which nutritive cells had become differentiated structures. The Bonnemaisoniales and Chaetangiaceae had a separate origin from Nemalionales in which nutritive auxiliary cells developed into generative auxiliary cells.

Later Hommersand discussed evolution within the Ceramiales, which he regards as evolved from the Nemalionales parallel with the Crypto-nemiales and Gigartinales. It seems therefore that evolution within the Florideophycideae may have progressed along a number of independent lines. Within the Ceramiales Hommersand suggests several different lines of evolution.

LIFE HISTORIES

When gametes fuse to form a zygote the chromosome number is doubled and before gametes can again be produced the original haploid number must be restored. In some algae this occurs when the zygote germinates; in other algae a diploid phase is introduced and from such a phase a haploid spore is produced.

In certain algae either phase is capable of reproducing itself and not giving rise to the alternate generation; hence the use of the term life history rather than life cycle. In those algae where there is only a diploid generation reduction division occurs at gamete formation and fusion of the gametes restores the double number of chromosomes.

In his studies on the Rhodophyceae, Svedelius coined a number of terms which have subsequently come into common use for the algae:

Haplont: A sexual haploid plant with only the zygote diploid *(Volvox).*

Diplont: A sexual diploid plant in which only the gametes are haploid, e.g. *Codium.*

Haplobiont: A plant possessing cytologically only one kind of individual in the life cycle: this may be either a haplont or a diplont.

Diplobiont: A plant possessing cytologically two kinds of individuals in the life cycle, and hence an alternation of generations.

Feldmann subsequently introduced the terms mono-, di- and trigenetic to describe the number of different independent morphological generations. In a later reconsideration of life histories Chapman and Chapman proposed a scheme.

A study of life histories within the Chlorophyceae shows that there are four principal types:

1. A unicellular or multicellular haploid generation with only the zygote diploid *(Chlamydomonas, Volvox).*
2. An alternation between morphologically identical diploid and haploid generations *(Ulva, Enteromorpha, Cladophord).*
3. An alternation between morphologically dissimilar generations, usually microscopic sporophytic (diploid) and macroscopic gametophytic (haploid) generation, e.g. *Ulothrix, Urospora, Acrosiphonia.* In the case of *Haticystis-Derbesia* neither generation could be considered microscopic).

4. A multicellular or coenocytic diploid generation in which the gametophyte has been eliminated, e.g. Siphonales.

If the haploid plants were primitively monoecious the first stage in evolution would presumably have been the development of the dioecious condition; this could have been followed by the intercalation of a sporophyte generation through delay in occurrence of the reduction division (e.g. *Monostroma* species with 'Codiolum' zygotic phase) leading to alternation of two generations.

Feldmann has put forward the view that the primitive life cycle in the Chlorophyta, Phaeophyta and Rhodophyta is the diplohaplophasic one but this is difficult to accept for the Chlorophyta because there are so many examples among the primitive orders, e.g. Volvocales, Chlorococcales.

A study of the life histories of the Phaeophyceae indicates that evolutionary progression in general conforms to evolutionary morphological progression. In most types the life history is not simple, because each generation is capable of repeating itself. Underlying the life history, however, is the fundamental nuclear cycle, but despite this there is no real regular alternation of generations except in the Dictyotales, Laminariales and some members of the Chordariales and Sphacelariales. It is evident that the primitive

Phaeophycean was either a plant with only a haploid generation in the life history or one in which there was isomorphic alternation. Very few members of the Ectocarpaceae are known to be haplonts, and in one good case, *E. virescens*, the parthenogenetic development of the eggs suggests degeneration rather than a primitive condition. The general weight of evidence would appear to be in favour of isomorphic alternation as the primitive Phaeophycean condition.

The work of Chihara and Magne and the French school has stimulated a dramatic revision of the whole problem of Rhodophycean life histories and their phylogeny. Although we now have a better appreciation of the situation than previously it is certain that even the present views will undergo further modifications as new work is carried out on the more primitive members.

The most primitive life history known at present is that

of *Cyanidium caldarium*, which reproduces by means of endospores only, so that it is haplophasic and monogenic. A more advanced condition is to be seen in *Erythrotrichia*, also haplophasic and monogenic but with male and female plants. Within the Bangiophycideae the most advanced type of life history is represented by *Porphyra* where it is haplophasic and heteromorphic.

Within the Floridеophycidae it appears that a truly haplophasic heteromorphic life history is very rare, but it does seem to exist in *Naccaria wiggii* and *Atractophora hypnoides*. If the higher red algae originated from the Bangiophycideae then these two species must be regarded as representing the primitive life history. The other alternative is to follow Feldmann and accept the diplohaplophasic life history as primitive.

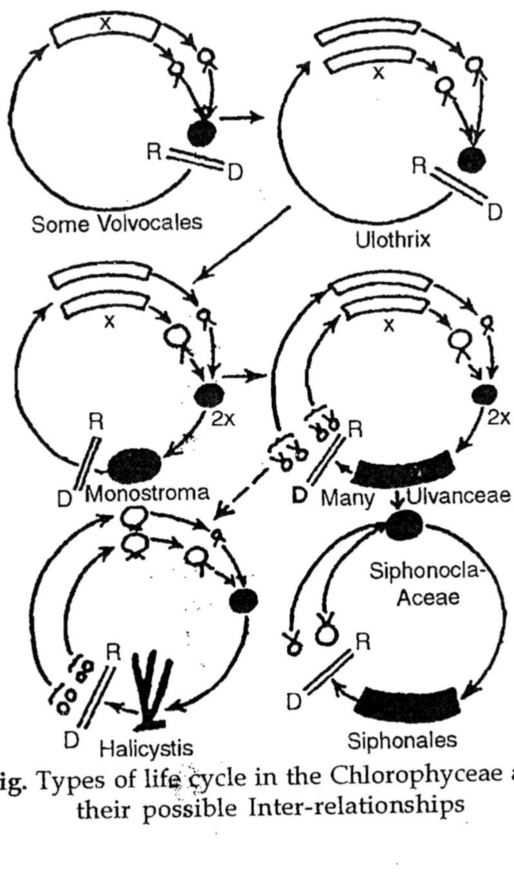

Fig. Types of life cycle in the Chlorophyceae and their possible Inter-relationships

(RD = position of reduction division in the life cycle).

The intercalation of a sporophytic generation, possibly first the diploid carposporophyte, followed by the dwarf tetrasporophyte, led to the development of the typical life history of the diplohaplophasic trigenetic type. *Liagora tetrasporifera* and other species with a similar life history represent forms in which there is only the diploid carposporophyte, though Magne prefers to treat these species as reduction forms in which the tetrasporophyte has become lost.

From the basic diplohaplophasic trigenic life history various evolutionary developments have taken place. One is the continuation of the heteromorphic alternation characteristic of so many of the Bonnemaisoniales, another has resulted in the reduction of the gametophyte generation *(Rhodothamniella floriduld)*, while in *Lomentaria rosea* in European waters and in the genera *Hildenbrandia* and *Apophloea* the gametophyte generation has become wholly suppressed.

Another development, which must be interpreted as retrogressive, can be seen in *Phyllophora membranifolia*, where the tetraspores are grouped into nemathecia on the diploid plant; in *P. brodiaei*, where the diploid phase disappeared and the namathecia can be regarded as parasites in the haploid thallus; and finally in *Ahnfeldtia* where meiosis no longer takes place and the nemathecia only produce monospores. Magne prefers to regard these last two examples as representing separate lines of reduction and there is clearly scope here for further study.

Feldmann has argued that in all three classes the primitive life history is the diplohaplophasic one with i$omorphic generations. The arguments against this view have been put forward in the case of the Chloro-phyceae and the thesis has been supported in the case of the Phaeophyceae. Feldmann's proposed primitive life history for the Rhodophyceae seems unnecessarily complex and is not supported by any primitive living representatives. It is evident that within the red algae further life history studies could lead to new interpretations of their phylogeny.

Within the various classes and divisions a number of definite morphological tendencies can be recognised, and these are repeated within the groups. In view of the plasticity of primitive organisms it is not surprising that the various potential lines of development have been exploited more than once.

14
Geographical Distribution of Algae

GEOGRAPHICAL DISTRIBUTION

The problem of algal biogeography is difficult because of the unreliability of earlier records and the somewhat scanty literature, especially for tropical and sub-tropical areas. Those studies that have been published have established certain general features which are summarized below:

1. There is a general resemblance between the algal floras of the West Indies and the Indo-Pacific. Vicarious pairs of species (i.e. two separate species closely related taxonomically and yet widely separated geographically) are known and even vicarious generic groups. Examples are the several vicarious pairs in the Dasycladaceous genus *Neomeris*. Murray's suggestion that such discontinuities were caused by major changes of climate in former epochs, would only appear to explain some cases, e.g. certain species in the Laminariaceae, but it is equally obvious that the present factors do not provide an adequate explanation. It seems best to postulate migration during an earlier epoch when there was a sea passage through the Panama isthmus, and this involves a migration not later than the Cretaceous.
2. There are some species which are common only to the Western Atlantic and the western part of the Indian Ocean around Madagascar, e.g. *Chamaedoris peniculwn* and three species of *Cladocephalus*.

Although at present there is no very adequate explanation for this distribution, three possible hypotheses may be suggested:

a. Migration via the Cape.
b. Migration via the Pacific and Panama, the

related species perhaps still existing in the Pacific but not yet recorded.

c. The related species or representatives in the interzone have died.

3. There are some genera common to the Mediterranean and the Indo-Pacific region, e.g. *Codium bursa* group, the vicarious pair *Halimeda tuna* in the Mediterranean and *H. cuneata* in the Indo-Pacific, *Acetabularia mediterranea* and other species *of Acetabularia* in the Indo-Pacific.

In this case the only satisfactory explanation is the existence of a former sea passage across the Suez isthmus. Species, e.g. *Caulerpa racemosa, Soliera dura,* are in fact known from the eastern Mediterranean, which must have migrated recently from the Indian Ocean via the Suez Canal. In the flora of the northern part of the Arabian Sea, out of a total of 137 species and varieties, 22 per cent are endemic, 52 per cent are Indo-Pacific and 60 per cent also occur in the Mediterranean and Atlantic Ocean, the most striking example being *Cystoclonium purpureum* which does not now exist between its widely separated stations along the southern shores of France and in the northern portion of the Arabian Sea.

4. The Indo-Pacific region is more probable as the source of various tropical and subtropical genera which can be classified into:
   a. genera with no Atlantic representatives.
   b. Genera with a few Atlantic species, e.g. *Halimeda, Caulerpa, Sargassum, Dictyota, Scinaia, Galaxaura.*

The following genera are probably of Atlantic origin: *Dasycladus, Pent- cillus, Cladocephalus, Batophora.*

5. Several families in the Laminariales, e.g. Laminariaceae, Alariaceae, are of Boreal Atlantic-Pacific discontinuity. These families must formerly have had a circum-arctic distribution but were pushed south by the onset of the Ice Age and then remained in their new habitat when the ice retreated. In this case, change of climate in a former epoch provides a satisfactory explanation of the present discontinuity.

Other genera, however, e.g. *Lessonia, Macrocystis, Ecklonia,* are of Antipodes-Northern Pacific discontinuity; *Macrocystis* in particular is primarily circum-antarctic, after which it is absent from the tropics, to reappear again on the Pacific coast of North America and around the shores of South Africa. The two southern species of *Macrocystis* appear to be identical with the two species in the northern hemisphere so that presumably they have disappeared from the intervening warm zone. Again, it must be concluded that their migration took place at a time when the temperatures of the ocean waters were more equable, unless it is assumed that the species have since become less tolerant towards temperature.

The existence in the Californian Miocene of extinct members allied to the extant southern genus *Cystophora* indicates a generic distribution almost identical with that of *Macrocystis.* Migration must have occurred from the south when temperatures were more equable and the species then disappeared with a rise in sea temperature. The absence of *Macrocystis* remains in this Miocene flora suggests that, if it came from the southern oceans, it must have migrated in post Miocene times.

6. Stephenson has shown how there are different elements in South Africa based on temperature, the cold Antarctic current on the west being responsible for the presence *ofEcklonia, Laminaria pdlida* and *Macrocystis.* A similar cold upwelling on the coast of California enables cold-water kelps to grow much further south than they do at corresponding latitudes on the east coast. By contrast the impact of the Gulf Stream on Great Britain enables a number of Mediterranean and other warm-water genera (e.g. *Cystoseird)* to occur in south-west Ireland and on the coast of Cornwall.

In the southern hemisphere Bennett and Pope have studied the biogeographical provinces of temperate Australia and they consider that four components can be recognized, though more may prove to be desirable.

a. Tropical, the main species occurring north of 26° south;

b. Warm-temperate;

c. Cool-temperate;

d. Universal.

They suggested that the average winter sea temperature of 10° C, which Stephenson regarded as the limiting value between warm and cool temperate, may be too low as the Victorian flora is definitely cool temperate and the average winter temperature is 11-8°C. The marine fauna and flora of Australia as a whole fall into a number of provinces which agree very closely with those established by Ferguson-Wood in his study of Australian dino-flagellates. In adjacent New Zealand there is a strong endemic element but Australian and sub-Antarctic components are also represented, the latter being found also in South Africa.

A recent review of the biogeographical provinces of Australia and New Zealand, based upon studies of animals, benthic algae and plank-tonic algae, has provided a much better understanding of the relationships in this part of the world with modifications of -earlier ideas. Knox introduced a transitional warm temperate region but there seems no case for treating these areas as other than warm temperate provided a subtropical zone is split off from the tropical belt. On this basis the Peronian province of Australia and the Kermadecian province of New Zealand would be subtropical. It is also evident that the provinces are related broadly to currents and temperatures of the main water masses. It is probable that further work on the subantarctic islands is necessary before one can finally say that the Antipodean and Kerguelan provinces are justified.

7. The peculiar distribution of certain algae in the southern hemisphere is significant in relation to Wegener's theory of Continental Drift. The presence of *Macrocystis, Splachnidium, Ecklonia* and *Ballia* in South Africa, widely distant from both Australia or South America, could conveniently be explained if at one time all three countries were part, with the Antarctic Continent, of a great southern continent which later broke up. If, instead, the permanence of

the present land masses and oceans is accepted, then it must be assumed that fertile fragments of these algae must have drifted to South Africa, presumably from South America via the Antarctic current. Portions of algae can be found floating in the ocean, though it is very doubtful whether fertile fragments could survive such a long journey. The most important factor concerned in biogeographical distribution is temperature, though locally presence or absence factors operate as well. The only other major feature affecting distribution is that of salinity which, with lowered values, yields special floras such as the one found in the Baltic Sea.

Setchell divided the surface waters of the oceans into zones according to the course of the 10°, 15°, 20° and 25° C isotherms, and showed that the great majority of algal species are confined to only one zone, a considerable number occur in two, only a small number occur in three zones, while the number extending over four or five zones are very few indeed and their distribution is usually by no means certain. In New England many of the species are apparently separated by the 20° C isotherm, which approximates closely to the position of Cape Cod, so that the flora to the north of the Cape is essentially different to that of the south.

Those species limited to one zone are called stenothermal while the wider-ranging forms are termed eurythermal. The former species are particularly characteristic of the warmer waters, but, even so, many apparent eurythermal species are found on examination to be essentially stenothermal. *Monostroma grevillei* and *Polysiphonia urceolata* are summer annuals in the cold waters of Greenland, but in the southern part of their range they develop in winter and early spring when the temperature will be the same as it is in the Greenland summer.

On the basis of these temperature studies Setchell recognised a number of climatic zones, e.g. Boreal, North Temperature etc., each of which contained species characteristic of the zone. An earlier study of Arctic algal floras by Borgesen and Jonsson indicated that the component species of the floras could be divided into five groups:

1. The Arctic group;
2. A sub-Arctic group;
3. Boreal Arctic group;
4. A cold boreal group;
5. A warm boreal group.

Although Iceland is so far north, the flora is nevertheless predominantly boreal because 54 per cent belongs to the last three groups. If the different districts of Iceland are compared with neighbouring floras it is extremely interesting to see how the floras of the various parts of the Icelandic coast show resemblances of widely separated areas.

A comparable analysis, in this classification, is Boreal and Austral, both of which represent warm temperate components. This classification has since been used as a basis for the analysis of the marine algal flora of Jamaica which showed that the principal element is the Caribbean, but, that while most of the remaining flora is tropical to warm temperate, there are also some cold temperate (Cold Boreal) species.

Studies in New Zealand have brought to light interesting distribution problems. It has been known for a long time that up to five distribution categories can be recognized:

1. Species not extending much north of Dunedin in South Island, e.g. *Pachymenia lusoria, Apophloea lyallii.*
2. Southern genera extending up to Kaikoura or Cook Strait, e.g.*Marginariella, Desmarestia.*
3. Species only north of E. Cape, e.g. *Halopteris hordacea, Carpophyllum angustifolium.*
4. Northern species extending south to the Cook Strait areas, e.g. *Caulerpa sedoides, Apophloea Sinclair ii.*
5. Universal component, e.g. *Splachnidium, Ecklonia.*

COMPARISON OF ENDEMIC GENERA IN AUSTRALIA AND NEW ZEALAND

In both regions it will be seen that there is a very high proportion of endemic species among the red algae.

Another interesting phenomenon that merits further study is the paucity of Phaeophyta in island groups in relation to Chlorophyta and Rhodophyta. Exactly why brown algae should be favoured by Continental or large island conditions is not immediately obvious.

LIFE FORM

A study of life form has proved useful in the case of terrestrial phanerogams and there would seem to be justification for some sort of algal life form classification comparable to that of Raunkiaer's for the flowering plants. Such a system can be used to give a quantitative picture of the composition of the vegetation and thus raise the problem aslo why some types are absent or reduced. Biological spectra, similar to those employed by Raunkiaer, form a convenient way, if used with caution, of comparing floras from two different areas although they are subject to the limitation that they do not indicate the dominant types unless frequency figures are available.

Oltmann's scheme of 1905, which is one of the earliest, was based largely upon morphological criteria, as also was that of Funk's which applied particularly to the algae of the Gulf of Naples. He distinguished four major primary groups which are not readily separated, and each was subdivided on a morphological basis.

Gislen in 1930 proposed yet another classification to include both plants and animals, the biological types referable to the plants being as follows:

Crustida (Crustaceous thallus):

1. Encrustida or encrusting forms, e.g. *Lithothamnion*
2. Torida or small cushion, e.g. *Rivularia*

Corallida (lime skeleton more or less developed):

1. Dendrida or tree-like forms, e.g. *Corallina*
2. Phyllida or leaf-like forms, e.g. *Udotea*
3. Umbraculida or umbrella-like forms, e.g. *Acetabularia*

Silvida (no lime skeleton):

1. Magncsilvida, or forms more than 10 cm high and with branches more than 1 mm thick.
   a. Graminida, e.g. *Zostera* (a phanerogam)
   b. Foliida, e.g. *Laminaria*
   c. Sack-form, e.g. *Enteromorpha*
   d. Palm form, e.g. *Lessonia*

e. Buoy form, e.g. *Nereocystis.*
f. Cord form, e.g. *Himanthalia*
g. Shrub-like form, e.g. *Chordaria*
h. *Sargassum* form
i. *Caulerpa* form

2. Parvosilvida (small delicate forms less than 100 mm high).

Setchell propounded a scheme based primarily on the conditions found in tropical waters, with particular reference to coral reefs. For this reason the classification is restricted because it would require considerable extension if the flora of colder waters were to be included, but its basis was largely ecological:

Heliophobes:

1. *Pholadophytes.* Forms nestling in hollows and avoiding excess light.
2. *Skiarophytes.* Forms growing under rocks or in their shade.

Heliophiles:

1. Metarrheophytes or attached flexible forms growing in moving water.
2. Lepyrodophytes or encrusting forms.
3. Herpophytes, composed of small creeping algae.
4. Tranophytes or boring species.
5. Cumatophytes or 'surf-loving' species.
6. Chordophytes, where the thallus has the form of a cord.
7. Lithakophytes or lime-encrusted species.
8. Epiphytes.
9. Endophytes.

The most workable scheme, so far, seems to be one proposed by Feldmann and modified later which is based on the same criteria, duration and perennation, that Raunkiaer used for the higher plants. The scheme is summarized as follows:

Annuals

1. Species found throughout the year. Spores or oospores germinate immediately.

EPHEMEROPHYCEAE: CLADOPHORA

2. Species found during one part of the year only.
   a. Algae present during the rest of the year as a microscopic thallus.

ECLIPSIOPHYCEAE

1. With microscopic gametophyte, *Sporochnus*.
2. With plethysmothallus, *Asperococcus*.
3. With microscopic sporophyte, *Ulothrix zonata*.
   a. Algae passing the unfavourable season in a resting stage.

HYPNOPHYCEAE — RESTING STAGE:

1. Oospores, *Vaucheria*.
2. Hormogones, *Rivularia*.
3. Akinetes, *Ulothrixpseudoflacca*.
4. Spores germinate and then become quiescent, *Dudresnaya*.

PERENNIALS

1. Frond entire throughout the year.
   a. Frond erect. Phanerophyceae: *Codium tomentosum*.
   b. Frond a crust. Chamaephyceae: *Hildenbrandia*.
2. Only a portion of the frond persisting the whole year.
   a. Part of the erect frond disappears.

HEMIPHANEROPHYCEAE: CYSTOSEIRA.

1. Basal portion of thallus persists. Hemicryptophyceae.
   a. Basal portion a disc, *Cladostephus*.
   b. Basal portion composed of creeping filaments, *Acetabularia*.

This scheme is a great advance on the earlier ones but it does not take adequate account of the effect of environment and, furthermore, being primarily of use for the marine algae, it does not take into consideration the numerous freshwater and terrestrial species.

The real test will come if and when this scheme is employed to give biological spectra. If the spectra from different localities, e.g. temperate and tropical regions, show a distinct difference then it should prove possible to extend its use as a means of comparing the vegetation from different regions. Such differences may be expected to open up problems, the solutions to which should yield us valuable information concerning the general biology and ecology of the species concerned.

ECOLOGICAL FACTORS

Any attempt to understand and interpret the zonation of a rocky coast or on a salt marsh must depend upon the part played by the different ecological factors. More than one factor will normally exert a controlling influence, but at one time of the year or at a certain phase in the life history of an organism one single factor may become paramount. A proper appreciation of some of the factors cannot be divorced from a study of the physiology of the algae and certain aspects of algal physiology must therefore be discussed in this connection.

The essential factors can be divided broadly into four groups:

1. Physiographic,
2. Physical,
3. Chemical,
4. Biological. Similarly, at any place it is possible to segregate the factors into those that are:
    a. Causal, i.e. responsible, so far as one can see, for the basic zonation;
    b. Presence or absence, i.e. factors such as freshwater or the type of rock which determine whether a species shall be present or not, but do not affect the level at which the species grows;
    c. Modificatory, i.e. factors such as spray, cracks, shade which cause a variation in the level normally occupied.

PHYSIOGRAPHIC FACTORS

The two important factors here are topography and the tides. The former is important because the nature of the coastline and the off-shore sea bed determine the degree of shelter or exposure to wave action. Waves themselves are determined by the stretch of water over which they are generated and by the strength of the winds. In some places it is the prevailing winds that will be most significant, e.g. trade-wind areas, while in other places it may be the frequency and force of gale winds.

Waves can be generated far out at sea and result in a major swell when local conditions are quite calm, and this kind of swell is not uncommon on Atlantic and Pacific coasts. The flatness or steepness of the shore is also significant, the former giving relatively calmer conditions than the latter. The presence of immediate off-shore submarine canyons, such as that of La Jolla in California, can also bring about local modifications of sublittoral conditions.

PHYSICAL DEGREE OF SHELTER OR EXPOSURE TO THE WAVES

This operates as a presence and absence factor, some species, *Durvillea, Postelsia,* possessing holdfasts that enable them to withstand strong wave action, or else it operates as a modificatory factor, as in elevating belts on exposed shores.

Even in the sublittoral there may be an effect, such as cover degree, where the maximum cover occurs at different levels under the changing conditions; thus, with shelter the level is at 1 m, with moderate wave action at 2 m, and on exposed coasts between 4-6 m. Under very exposed conditions it is evident that the sporelings of the algae must be able to withstand severe wave action, and the attachment of spore or zygote to the rock surface must take place very rapidly.

In the case of Laminarian species, e.g. *L. sinclairii, Postelsia,* the ovum must either be fertilized *in situ* or else there must be rapid adherence the moment a fertilized egg touches a rock surface. This factor will also operate on large freshwater lakes where the waves can be quite large. Such waves may indeed determine the type of algal community at the lake edge, e.g. where there is exposure to wave action Cyanophyceae tend to predominate, whereas in more sheltered places Chlorophyceae can play an important role.

WAVE ACTION AND TIDES

By wave action is meant the actual height of the waves and the degree of splash and sprays those results from them. This is essentially a modifying factor, elevating the zones, especially the upper ones, often to a considerable degree.

Certain basic features of the tides are as follows:

1. The maximum tidal range varies from place to place.
2. The tidal rhythm can vary from place to place so that it is possible to observe the usual twice-daily tides (that are generally of unequal height) or the daily but less-common single high and low tides (diurnal).
3. The time of low water of spring tides varies from place to place. If it occurs around mid-day, especially in summer, this could have a profound effect upon any species that cannot tolerate high temperatures and consequent desiccation, or those in which photosynthesis is inhibited by high light intensities during exposure.

Tidal Range

This is a modifying factor as it is primarily responsible for the actual width of the algal zones. Even with a very small tidal range, e.g. Caribbean, it is possible to observe a distinct zonation of the organisms. This factor also determines the extent to which rock pools may exist, the greater the tidal range the greater the variety among the rock pools. The different types of tidal variation are of great importance and may alter the position of an organism on the shore.

Tidal Currents

In certain places there may be very pronounced currents close inshore where algae may grow, e.g. between Orkney Is. and the north of Scotland, at the mouth of the Bay of Fundy, etc. Such currents affect the growth of the plants but they can also operate as a presence or absence factor.

Thus on the east coast of Scotland the strength of the current determines whether *Lamin-aria saccharina* or *L. hyperborea* is the dominant. In New Zealand increased current strength seems responsible for the presence of the so-called *var. richardiana ofEcklonia radiata,* which is therefore probably only an ecological growth form. The strength of the current flowing up salt marsh and mangrove swamp creeks determines the degree of erosion, and hence the extent to which creek bank algal communities can develop.

Silt Load

In some areas, e.g. near the mouth of the River Severn in England, Bay of Fundy in Nova Scotia, the load of silt carried by the sea exerts a distinct effect upon the fauna and flora. Certain organisms are unable to tolerate silt deposition, in the case of algae this may be due to inhibition of photosynthesis. Silt load can therefore operate as a presence or absence factor.

Inter-tidal Exposure

This is the exposure suffered at any level between two consecutive tides and it varies with the type of tide, i.e. spring or neap. Exposure operates indirectly insofar as it affects water loss, temperature changes of the thallus (especially in tropics), and salinity of the cells. The degree of atmospheric humidity is also important during the periods of exposure, and it is this, together with the high temperature in the drier areas of the tropics that prevents algae from occupying the shore but enables them to form a dense covering at high moist latitudes.

The length of the inter-tidal exposure also affects the conditions in rock pools and salt pans, the diurnal changes being greater in the higher pools and pans where the length of exposure is longer. With one type of tide there will be certain levels on the shore where the inter-tidal exposure can be doubled, e.g. at 3-5 ft. Similarly, the submergence period may be drastically altered over a short vertical range, e.g. just above 1-5 ft it is about 10 hr, whereas just below 1.5 ft it is about 30 hr. The effect of this phenomenon in relation to critical levels may well be profound.

Continuous Exposure and Submergence

At levels lying between lowest high water of neap tides

and extreme highest water of spring tides there will be periods of two or more days during the neap tides when the level will not be inundated at all. Similarly, between highest low water of neap tides and extreme lowest low water of spring tides there will be days during the neap tides when the levels will be continuously submerged. In the former case, desiccation, temperature and salinity changes will be far more profound than those that occur between two consecutive tides; in the latter case continual submergence may affect the metabolic activities and the compensation point, though this will depend upon the maximum tidal range.

It is probable that the periods of continuous exposure are of far greater importance than inter-tidal periods, especially during the sporeling stages. Gail has suggested that it is the desiccation of young plants that prevents the appearance of algae outside their usual zones, and it is a remarkable fact that sporelings of fucoids are usually very strictly confined from an early stage to the level they occupy as adult plants.

A rather different exposure phenomenon can be observed around large lakes. Evaporation in the summer very often lowers the water level so that the winter and summer levels are different. This zone of exposure between the two seasonal levels may be occupied by a special algal community, or it may form a bare zone between the winter water-edge community and the lower, summer water-edge community.

Diurnal variations in temperature may affect the metabolic activities of the algae and extreme variations in salinity can also bring about the death of some species. Water loss during periods of exposure is quite clearly an important factor. On salt marshes it is undoubtedly responsible for the occurrence of gelatinous Cyanophyceae at the higher levels where the exposure periods are long. Among others, Pringsheim, Zaneveld, Isaac, Kanwisher, Brown and Johnson and Chapman have studied rate of water loss in intertidal algae. Much of the work has centered around the principal fucoids found on European shores.

In *Pelvetia canaliculata* the main water loss occurs in the first six hours of exposure while in *Fucus spiralis* var. *platycarpus, F. vesiculosus, F. serratus* and *Ascophyllum* the maximum water loss can be spread over 18 h: in *F. vesiculosus* as much as 90 per cent of the total initial water can be lost in 1-5 h. *Fucus spiralis* var. *platycarpus* loses its water at the slowest rate and it occurs at the highest level on the shore. Haas and Hill have also shown that the higher the alga grows the greater is the fat content.

This fat is largely contained in the cell walls and examination has shown that the thickness of the cell wall is greater in plants growing higher on the shore.

These cell walls decrease in thickness when subjected to desiccating conditions, so it must be assumed that a large part of the water lost is contained there. Those species which lose water most slowly also reabsorb it most slowly and, as a result, the growth rate of the highest species therefore tends to be the slowest. It would appear that an important factor controlling zonation, so far as these fucoids are concerned, is the biochemical nature and properties of the cell wall.

In the fucoid *Hormosira* the rate of water loss is correlated with bladder size, the larger the bladder the slower the rate of water loss. Plants with large bladders occur at the higher levels on the shore so that morphology provides protection in this species.

For some years now detailed studies have been made of water loss from selected New Zealand inter-tidal algal under varying conditions of exposure. In all cases the water loss is related to the maximum exposure time.

The probable water loss is that which is estimated to occur under summer conditions and it is believed this is the principal causal factor determining upper limits. With *Caloglossa* the high humidity in the mangrove swamps and in rock cracks reduces water loss to a level that the delicate thalli can tolerate. The high humidity of the mangrove swamps also reduces water loss in *Hormosira*. In the case of the two species of *Carpophyllum* successful transplants of adult plants to higher levels suggests that it is water loss from the sporelings that determines their upper limit.

A further feature that has received very little attention is the effect of desiccation upon the major metabolic processes, respiration and photosynthesis. A study upon *Hormosira* has shown that the effects may be quite different between sporeling and adult. This may be of profound significance and further work is clearly required. In *Scytothamnus* respiration decreases with increasing desiccation and lowered humidity.

During exposure humidity may be very important. With humidities of 80 per cent r.h. or more and with a water loss not exceeding 40 per cent *Hormosira* will photosynthesise for up to 3 h exposure, but the degree of photosynthesis is dependent upon light intensity as well.

RELATIONSHIP BETWEEN LIGHT INTENSITY AND PHOTOSYNTHESIS

Similarly with the two species of *Carpophyllum* there is some photosynthesis at high humidities during exposure. *Scytothamnus,* on the other hand, shows assimilatory balance only at 100 per cent r.h. and high light intensities. In *Caloglossa* there is no photosynthesis below 60 per cent r.h. or a desiccation value of 35 per cent.

Relatively little study has been made of the effect of subsequent re-immersion upon photosynthesis and respiration. In the case of adult *Hormosira* the immediate effect of submergence is to cause a rise in photosynthesis. This rise is dependent upon;

1. The ecological form of the algae,
2. Degree of desiccation,
3. Intensity of illumination.

With *Caloglossa leprieurii* photosynthesis on resubmergence at 2000 ft-candle decreased with increasing desiccation up to 40 per cent water loss after which photosynthesis became disorganized. At lower light intensities a greater degree of water loss could be tolerated. *Gelidium caulacantheum* continues to photosynthesize after 8 h exposure and 50 per cent water loss. *Scytothamnus australis* shows a decrease in photosynthesis on re-immersion, but even up to 70 per cent desiccation the decrease is not more than half that of the control.

The water relations of fucoids and other large algae may be compared with the *Enteromorpha* and Chrysophyceae belts of the Dover cliffs. Anand found that the *Enteromorpha* mat lost 25 per cent of its moisture in the first 3 h of exposure, whilst the Chrysophyceae belt lost 8-4 per cent.

The measurements were obtained by the simple method of weighing portions of the mat cut out so that they fitted into waterproof paper dishes which could be put back into position on the shore. The reduced loss from the Chrysophyceae belt as compared with the *Enteromorpha* is due to the gelatinous nature of the Chrysophyceae.

PHYSICAL FACTORS

Substrate

The nature of the substrate, whether solid rock, boulders, pebbles, sand, mud or peat, is of fundamental importance in anchoring the plant and the general aspect of the flora is determined to a large extent by this factor. It is essentially a presence or absence factor. On rocky coasts the angle of slope and the presence of cracks may affect the local occurrence of some species, either operating as a presence or absence factor or as a modifying factor. The geclogical nature of the rock may also be important, e.g. soft rocks such as sandstone rarely carry large algae.

Pressure

Normally this is not an important factor, but Damant has shown that in the case of vesicular fucoids there is a limit to the depth at which bladders will exist because the gas is forced out by the pressure of the water forces. This depth will therefore mark the approximate lower limit to which the species can descend, since the vesicles probably represent a flotation mechanism and keep the thallus near the surface when submerged.

A similar phenomenon has been observed in the case of *Egregia* where the shape of the bladders appears to be related to water depth, the deeper plants having spherical bladders and the shallow plants pear-shaped ones. In *Ascophyllum* and *Fucus,* Aleem has shown that the pressure is related to oxygen production and it therefore increases during the day and decreases at night.

Illumination

There are problems here involving seasonal'variation in

light intensity, diurnal variations in relation to times of high water, and the actual light intensity and spectrum at different depths. Practically nothing is known about the first two effects, and it is impossible therefore to assess their importance. It can, however, be pointed out that differences in illumination intensity at different latitudes appear to determine the maximum depth to which algae can descend.

Thus in the North Sea the limit appears to be about 40-50 m, in the tropics it is about 100 m whilst in the clear waters of the Mediterranean it is about 130-180m. Many of the deep-growing algae (and possibly others) are fully light-saturated at quite low values. Blades of *Macrocystis* are fully saturated at 1000 ft-candles and parts of *Egregiasit* 1400 ft-candles.

The third effect is undoubtedly important because measurements show that the incident light is cut down very considerably at even a depth of 1 m, while at 2 m only about 25 per cent of the surface light can penetrate. Waters of different turbidity may absorb light quite differently and this in turn affects the relative photosynthetic rate and the compensation point level. During its passage through sea water the spectral composition of the light changes considerably, the red and yellow part of the spectrum being preferentially absorbed, such that the blue and green portions assume greater importance.

Algae living in the sublittoral fringe and the upper portion of the sublittoral are subject to very different conditions as compared to algae living higher on the shore, so far as photosynthesis is concerned. There is also the different effect of submergence and emergence upon photosynthesis, the process being affected not only by the light change but also by the moisture and temperature changes. Algae can be found at 100 m depth in the arctic where the light intensity is extremely low and it has been suggested that such algae must be heterotrophic.

It is also evident that temporary changes in light intensity as a result of cloudiness or chop on the water may affect photosynthesis. The outer edges of *Macrocystis* beds in California often occur at depths of 12-20 m in turbid water but at 27-5-30 m or more in clearer waters. Tschudy first showed that maximum photosynthesis occurred at the surface on choppy days but at about 5 m on calm days. Stocker and Holdheide used *Fucus spiralis* var. *platycarpus* and *F. serratus* where the difference between sunny and cloudy days was extremely pronounced.

It would seem that on exposure the fall off in assimilation rate on sunny days is correlated with water content because the exposed thalli quickly dry up and cease to assimilate. An investigation into photosynthesis of *Hormosira* under submerged and exposed conditions at Auckland produced similar results.

Work from Japan on sporelings has shown that except for *Monostroma,* sporeling growth in green algae is greatly retarded under decreased light, while that of red and brown algae tends to be increased. Sporelings of green algae also grew more rapidly under yellow, red or blue light than under green light. Length of the light period may also be involved. Thus *Enteromorpha* sporelings grow better under long-day conditions, whereas those of *Mono-stroma* grow best at first under medium day length but that later short days are optimal.

At low light intensities there is little difference in photosynthesis between adult and sporeling estuarine *Hormosira banksii.* With increasing light intensity the juvenile plants respond very much more rapidly than do the adults. *Carpophyllum plumosum* sporelings, on the other hand, have a higher photosynthetic rate than the adults at comparable light intensities. Maximum efficiency occurs around 1500 ft. candles and this suggests that the species is essentially a shade alga. With *Laminaria hyperborea* the gametophytes and sporelings are distinct shade plants with light saturation at while the adult sporophytes are not saturated.

There has been considerable discussion over many years in connection with the pigment composition of the algae in relation to photosynthesis and depth. Quite early Engelmann put forward the hypothesis that the colour of an alga is complementary to that of the incident light. In the Chlorophyceae maximum rates of photosynthesis occur in the red and blue region of the spectrum, where maximum absorption takes place by the photosynthetic pigments. In sea water the red and much of the blue region are rapidly eliminated such that Chlorophyceae living at any depth would be expected to have a very low rate of photosynthesis, although certain siphonaceous green algae are commonly found in deeper waters.

In the Phaeophyceae there is greatly increased absorption in the green region due to the presence of fucoxanthin. As this carotenoid, unlike nearly all others, has a very high efficiency (100 per cent) of absorbed energy transfer to chlorophyll, there is consequently a much increased rate of photosynthesis, relative to that in the blue and red regions. Consequently for the Phaeophyceae, living at depths where the light spectrum is predominantly blue and green, photosynthesis is not the problem it is in the Chlorophyceae.

In the Rhodophyceae there is greatly increased light absorption in the green yellow regions due to the presence of the photoreactive biliproteins. These pigments transfer their absorbed energy to the chlorophyll so that light, not predominantly absorbed by chlorophyll, is utilized in photosynthesis.

In all algal photosynthetic studies consideration must be given to the absorption curves and the action spectra. The latter is the more important because it indicates the extent to which the different portions of the absorbed spectrum are utilized in photosynthesis. The absorption spectra of a green and red alga at different depths, and it will be seen that in *Monostroma* there is a rapid decrease in the absorption of the red wave lengths as depth increases.

All exhibit an optimum light intensity for assimilation, which in the case of the littoral algae is in the same region as that of cormophytic land plants. In sub-littoral algae, such as *Laminaria saccharina,* the optimum is at a much lower light intensity. Because certain algae are more efficient metabolically at low light intensities or when the spectral composition of the light has been modified, it has been suggested that it is possible to recognize sun and shade algae. While there may be a definite ecological value in the distinction, nevertheless further experimental work is desirable before any final conclusions are reached.

The compensation depth or light intensity at which photosynthesis just balances respiration may be very important. This differs for various species and is also dependent on temperature. At 15°C it is 15 ft-candles for *Macrocystis,* 38 ft-candles for *Fucus serratus* and 32 ft-candles for *Laminaria saccharina.* In the case of *Carpophyllum plumosum* in order to balance the respiratory loss during 13 h submergence in the dark at 13° C, photosynthesis over a period of 11 h at a light intensity of not less than 200 ft-candles is necessary.

Field light measurements indicate that this limiting light intensity (as an annual mean) is reached at levels where the species ceases to be present and this can be related to light penetrating different kinds of off-shore waters.

Apparent photosynthesis between juvenile, medium dissected and basal entire leaves of deep-water plants of *Egregia laevigata.* The same phenomenon, even more pronounced, is exhibited by different parts of *Macrocystis pyrifera.*

For this species it is quite evident that the rate of photosynthesis increases with age of the blade. In the case of *Egregia* it is the middle-aged dissected appendages that exhibit the highest rate. Some recent work on *Macrocystis pyrifera* in California has shown that the net photosynthesis varies not only with depth, as evidenced by the rates exhibited by blades from different positions on the plant, but also with temperature.

Ehrke suggested that there is a correlation between the temperature at which maximum assimilation takes place and the average temperature of the month of maximum development. The problem, however, is not as simple as this. Thus Lampe found that in winter the assimilation rate of *Fucus serratus* plants rises when it is measured in sunlight under conditions of increasing temperature; on the other hand, in the case of a red alga such as *Porphyra,* when the temperature is raised above 15° C the assimilation curve is lowered immediately in diffuse light and after seven days in sunlight.

This suggests that *Fucus* is an eurythermal species, tolerating a wide range of temperature, while *Porphyra* is stenothermal, tolerating a narrow range. Comparable results for *Fucus serratus* were obtained by Hyde, who found that between 15° C and 20° C the assimilation rate could be increased by raising the light intensity, and that there was a certain light value which yielded an optimum in the rate of assimilation.

After a period of exposure there is the problem of return to normal assimilation rate, because experiments show that the assimilation rate is markedly reduced during exposure. When reinundated, species such as *Viva linza* and *Porphyra umbilicalis* take up water at once, and very soon are assimilating at their normal rate. The members of the Fucaceae behave differently and their behaviour can be correlated directly with the level they occupy on the shore and the periods of exposure they undergo.

There would seem little doubt that illumination and its effect on photosynthesis is an important causal factor, and probably is largely responsible for determining the lower limits of a number of algae. It is, however, a very complex factor, and much further work is necessary before its operation is fully unravelled.

1. Sublittoral algae which are never exposed to the air; they are resistant to a concentration of 1.4 times that of sea water.
2. Algae of low-water mark and the lower littoral tide pools which never become dry; these are resistant up to a concentration of 2-2 times that of sea water,
3. Algae of the littoral belt; these are often completely exposed and they can resist a concentration of 3-0 times that of sea water.

The influence of lowered salinity upon metabolic processes requires further study, especially with respect to the depth compen-sation point.

Salinity changes also occur in pools and salt pans, especially those at high levels, during neap tidal periods when there may be considerable evaporation. This factor can operate as a presence or absence factor in high-level pools. In pools, intertidal algae exhibit a greater tolerance to changes in salinity than do sublittoral algae, and this may well explain why the latter are restricted to pools in the lower half of the littoral zone.

Substrate

Generally, differences in chemical composition have little

effect on the flora. An exception is the presence of chalk, the algal flora of chalk cliffs differing distinctly from that of metamorphic and igneous rocks.

PH (Acidity)

The pH of sea water normally ranges from about 7-9 to 8-3 and the factor is of no great importance for littoral algae though it might be of significance in some of the higher rock pools where, during the day, the pH may rise to 10. Most algae that have been investigated tolerate a pH range of 6-8-9-6 and the pH of pools is rarely outside this range.

Oxygen Content

The oxygen tension in sea water is normally low but sufficient for metabolic purposes. In pools, during the day the oxygen content may rise substantially.

Nutrients

This factor is of importance in controlling the seasonal periodicity of marine and freshwater plankton. It is also of importance in lakes where oligotrophic and eutrophic water differ chemically, and this difference is reflected in the vegetation. High-level rock pools well supplied with bird droppings commonly support a very rich flora of microscopic algae.

BIOLOGICAL FACTORS

Animals

Competition between plants and animals may in certain cases be sufficiently severe to eliminate some algae, it therefore operates as a presence or absence factor. If *Patella* is present in abundance it may eat up the *Entero-morpha* felt and so stop *Fucus* sporelings from becoming established. The attacks of boring mollusca are probably responsible for the eventual detachment of *Durvillea* plants, and in Europe *Helicion pellucidum* is said to bring about the detachment of *Laminaria saccharina* plants.

On salt marshes the mollusc *Hydrobia uhae* can cause havoc among beds of *Ulva*. Sea urchins, if present in abundance, can completely clear rock pools of nearly all algae and, in conjunction with other factors, may help to set the lower limit, e.g. grazing of *Carpophyllwn plumoswn* by *Evechinus chloroticus* and *Lunella smaragda* in New Zealand probably helps to determine the lower limit in the clearer waters. Removal *of Echinus esculentus* increases the density *of Laminaria hyperborea* from 5-1/m to 22-7/m^2. Operating in the other direction is the occurrence of the red alga *Pachymenia himantophora* in relation to the prior presence of barnacles.

Plants

Many of the possible relationships here have not been worked out. There are, of course, the host-parasite and host-epiphyte relationships, because some of the epiphytes appear to be restricted in their host requirements. The establishment of *Fucus* sporelings also seems to depend very largely upon the pre-development of a green algal felt whilst the establishment of *Porphyra tenera* sporelings on the bamboo poles or nets in Japanese harbours is dependent upon the development of a diatomaceous film.

In the case of ®epiphytes certain species at least may be restricted to specific parts of large host plants. Such epiphytes can be segregated into epithallic epicaulic (stem), epiphyllic ('leaf'), epirhizoidic,.